TRIUMPH GT6 1966-1974

Compiled by
R.M. Clarke

ISBN 0 907073 21 2

Distributed by
Brooklands Book Distribution Ltd.
'Holmerise', Seven Hills Road,
Cobham, Surrey, England

BROOKLANDS BOOKS SERIES

AC Ace & Aceca 1953-1983
AC Cobra 1962-1969
Alfa Romeo Giulia Berlinas 1962-1976
Alfa Romeo Giulia Coupés 1963-1976
Alfa Romeo Spider 1966-1987
Aston Martin Gold Portfolio 1972-1985
Austin Seven 1922-1982
Austin A30 & A35 1951-1962
Austin Healey 100 1952-1959
Austin Healey 3000 1959 1967
Austin Healey 100 & 3000 Collection No. 1
Austin Healey 'Frogeye' Sprite Collection No. 1
Austin Healey Sprite 1958-1971
Avanti 1962-1983
BMW Six Cylinder Coupés 1969-1975
BMW 1600 Collection No. 1
BMW 2002 1968-1976
Bristol Cars Gold Portfolio 1946-1985
Buick Riviera 1963-1978
Cadillac Automobiles 1949-1959
Cadillac Eldorado 1967 1978
Cadillac in the Sixties No. 1
Camaro 1966-1970
Chevrolet 1955-1957
Chevrolet Camaro Collection No. 1
Chevelle & SS 1964-1972
Chevy II Nova & SS 1962-1973
Chrysler 300 1955-1970
Citroen Traction Avant 1934-1957
Citroen 2CV 1949-1982
Cobras & Replicas 1962-1983
Cortina 1600E & GT 1967-1970
Corvair 1959 1968
Daimler Dart & V-8 250 1959-1969
Datsun 240z & 260z 1970-1977
De Tomaso Collection No. 1
Dodge Charger 1966-1974
Excalibur Collection No. 1
Ferrari Cars 1946-1956
Ferrari Cars 1962-1966
Ferrari Cars 1969-1973
Ferrari Dino 1965-1974
Ferrari Dino 308 1974 1979
Ferrari 308 & Mondial 1980-1984
Ferrari Collection No. 1
Fiat X1/9 1972 1980
Ford Falcon 1960-1970
Ford Mustang 1964-1967
Ford Mustang 1967-1973
Ford RS Escort 1968 1980
High Performance Escorts MkI 1968-1974
High Performance Escorts MkII 1975-1980
Hudson & Railton Cars 1936-1940
Jaguar (& S.S) Cars 1931-1937
Jaguar Cars 1957-1961
Jaguar Cars 1961-1964
Jaguar Cars 1964-1968
Jaguar E-Type 1961-1966
Jaguar E-Type 1966-1971
Jaguar XKE Collection No. 1
Jaguar XJ6 1968-1972
Jaguar XJ6 Series II 1973-1979
Jaguar XJ6 & XJ12 Series III 1979-1985
Jaguar XJ12 1972 1980
Jaguar XJS 1975-1980
Jensen Cars 1946-1967
Jensen Cars 1967-1979
Jensen Interceptor Gold Portfolio 1966-1986
Lamborghini Cars 1964-1970
Lamborghini Cars 1970-1975
Lamborghini Countach Collection No. 1
Lamborghini Countach & Urraco 1974-1980
Lamborghini Countach & Jalpa 1980-1985
Lancia Stratos 1972 1985
Land Rover 1948-1973
Land Rover Series II & IIa 1958-1971
Land Rover Series III 1971-1985
Lotus Cortina 1963-1970
Lotus Elan 1962-1973
Lotus Elan Collection No. 1
Lotus Elan Collection No. 2
Lotus Elite 1957-1964
Lotus Elite & Eclat 1974-1981
Lotus Turbo Esprit 1980-1986
Lotus Europa 1966-1975
Lotus Europa Collection No. 1
Lotus Seven 1957-1980
Lotus Seven Collection No. 1
Maserati 1965-1970
Maserati 1970-1975
Mazda RX-7 Collection No. 1
Mercedes 230/250/280SL 1963-1971
Mercedes 350/450SL & SLC 1971-1980
Mercedes Benz Cars 1949-1954
Mercedes Benz Cars 1954-1957
Mercedes Benz Cars 1957-1961
Mercedes Benz Competition Cars 1950-1957
Metropolitan 1954-1962
MG Cars 1929-1934
MG TC 1945-1949
MG TD 1949-1953
MG TF 1953-1955
MG Cars 1957 1959
MG Cars 1959-1962
MG Midget 1961-1980
MG MGA 1955-1962
MGA Collection No. 1
MGB Roadsters 1962-1980
MGB GT 1965-1980
Mini Cooper 1961-1971
Morgan Cars 1960-1970
Morgan Cars 1969-1979
Morris Minor Collection No. 1
Old's Cutlass & 4-4-2 1964-1972
Oldsmobile Toronado 1966-1978
Opel GT 1968-1973
Pantera 1970-1973
Pantera & Mangusta 1969-1974
Plymouth Barracuda 1964-1974
Pontiac GTO 1964 1970
Pontiac Firebird 1967-1973
Pontiac Tempest & GTO 1961-1965
Porsche Cars 1960-1964
Porsche Cars 1964-1968
Porsche Cars 1968-1972
Porsche Cars in the Sixties
Porsche Cars 1972-1975
Porsche 356 1952-1965
Porsche 911 Collection No. 1
Porsche 911 Collection No. 2
Porsche 911 1965-1969
Porsche 911 1970-1972
Porsche 911 1973-1977
Porsche 911 Carrera 1973-1977
Porsche 911 SC 1978-1983
Porsche 911 Turbo 1975-1984
Porsche 914 1969-1975
Porsche 914 Collection No. 1
Porsche 924 1975-1981
Porsche 928 Collection No. 1
Porsche 944 1981-1985
Porsche Turbo Collection No. 1
Reliant Scimitar 1964-1986
Rolls Royce Silver Cloud 1955-1965
Rolls Royce Silver Shadow 1965-1980
Range Rover 1970-1981
Rover 3 & 3.5 Litre 1958-1973
Rover P4 1949-1959
Rover P4 1955-1964
Rover 2000 + 2200 1963-1977
Rover 3500 1968-1977
Rover 3500 & Vitesse 1976-1986
Saab Sonett Collection No. 1
Saab Turbo 1976-1983
Singer Sports Cars 1933-1934
Studebaker Hawks & Larks 1956-1963
Sunbeam Alpine & Tiger 1959-1967
Thunderbird 1955 1957
Thunderbird 1958-1963
Triumph 2000-2.5-2500 1963-1977
Triumph Spitfire 1962-1980
Triumph Spitfire Collection No. 1
Triumph Stag 1970-1980
Triumph Stag Collection No. 1
Triumph TR2 & TR3 1952 1960
Triumph TR4.TR5.TR250 1961-1968
Triumph TR6 1969-1976
Triumph TR6 Collection No. 1
Triumph TR7 & TR8 1975-1981
Triumph GT6 1966-1974
Triumph Vitesse & Herald 1959 1971
TVR 1960-1980
Volkswagen Cars 1936-1956
VW Beetle 1956-1977
VW Beetle Collection No. 1
VW Golf GTi 1976-1986
VW Karmann Ghia 1955-1982
VW Scirocco 1974-1981
Volvo 1800 1960-1973
Volvo 120 Series 1956-1970

BROOKLANDS MUSCLE CARS SERIES

American Motors Muscle Cars 1966-1970
Buick Muscle Cars 1965-1970
Camaro Muscle Cars 1966-1972
Capri Muscle Cars 1969-1983
Chevrolet Muscle Cars 1966-1972
Dodge Muscle Cars 1967-1970
Mercury Muscle Cars 1966-1971
Mini Muscle Cars 1961-1979
Mopar Muscle Cars 1964-1967
Mopar Muscle Cars 1968-1971
Mustang Muscle Cars 1967-1971
Shelby Mustang Muscle Cars 1965-1970
Oldsmobile Muscle Cars 1964-1970
Plymouth Muscle Cars 1966-1971
Pontiac Muscle Cars 1966-1972
Muscle Cars Compared 1966-1971
Muscle Cars Compared Book 2 1965-1971

BROOKLANDS ROAD & TRACK SERIES

Road & Track on Alfa Romeo 1949-1963
Road & Track on Alfa Romeo 1964-1970
Road & Track on Alfa Romeo 1971-1976
Road & Track on Alfa Romeo 1977 1984
Road & Track on Aston Martin 1962-1984
Road & Track on Audi 1952-1980
Road & Track on Audi 1980-1986
Road & Track on Austin Healey 1953-1970
Road & Track on BMW Cars 1966-1974
Road & Track on BMW Cars 1975-1978
Road & Track on BMW Cars 1979-1983
Road & Track on Cobra, Shelby & Ford GT40 1962-1983
Road & Track on Corvette 1953-1967
Road & Track on Corvette 1968-1982
Road & Track on Corvette 1982-1986
Road & Track on Datsun Z 1970-1983
Road & Track on Ferrari 1950-1968
Road & Track on Ferrari 1968-1974
Road & Track on Ferrari 1975-1981
Road & Track on Ferrari 1981-1984
Road & Track on Fiat Sports Cars 1968-1981
Road & Track on Jaguar 1950-1960
Road & Track on Jaguar 1961-1968
Road & Track on Jaguar 1968-1974
Road & Track on Jaguar 1974-1982
Road & Track on Lamborghini 1964-1985
Road & Track on Lotus 1972-1981
Road & Track on Maserati 1952-1974
Road & Track on Maserati 1975-1983
Road & Track on Mazda RX7 1978-1986
Road & Track on Mercedes Sports & GT Cars 1970-1980
Road & Track on MG Sports Cars 1949-1961
Road & Track on MG Sports Cars 1962 1980
Road & Track on Mustang 1964-1977
Road & Track on Peugeot 1955-1986
Road & Track on Pontiac 1960-1983
Road & Track on Porsche 1951-1967
Road & Track on Porsche 1968-1971
Road & Track on Porsche 1972-1975
Road & Track on Porsche 1975-1978
Road & Track on Porsche 1979-1982
Road & Track on Porsche 1982-1985
Road & Track on Rolls Royce & Bentley 1950-1965
Road & Track on Rolls Royce & Bentley 1966-1984
Road & Track on Saab 1955-1985
Road & Track on Toyota Sports & G T Cars 1966-1986
Road & Track on Triumph Sports Cars 1953-1967
Road & Track on Triumph Sports Cars 1967-1974
Road & Track on Triumph Sports Cars 1974-1982
Road & Track on Volkswagen 1951-1968
Road & Track on Volkswagen 1968-1978
Road & Track on Volkswagen 1978-1985
Road & Track on Volvo 1957-1974
Road & Track on Volvo 1975-1985

BROOKLANDS CAR AND DRIVER SERIES

Car and Driver on BMW 1955-1977
Car and Driver on BMW 1977-1985
Car and Driver on Cobra, Shelby & Ford GT40 1963-1984
Car and Driver on Datsun Z 1600 & 2000 1966-1984
Car and Driver on Corvette 1956-1967
Car and Driver on Corvette 1968-1977
Car and Driver on Corvette 1978-1982
Car and Driver on Ferrari 1955-1962
Car and Driver on Ferrari 1963-1975
Car and Driver on Ferrari 1976-1983
Car and Driver on Mopar 1956-1967
Car and Driver on Mopar 1968-1975
Car and Driver on Pontiac 1961-1975
Car and Driver on Porsche 1955-1962
Car and Driver on Porsche 1963-1970
Car and Driver on Porsche 1970-1976
Car and Driver on Porsche 1977-1981
Car and Driver on Porsche 1982-1986
Car and Driver on Saab 1956-1985
Car and Driver on Volvo 1955-1986

BROOKLANDS MOTOR & THOROUGHBRED & CLASSIC CAR SERIES

Motor & T & CC on Ferrari 1966-1976
Motor & T & CC on Ferrari 1976-1984
Motor & T & CC on Lotus 1979-1983
Motor & T & CC on Morris Minor 1948-1983

BROOKLANDS PRACTICAL CLASSICS SERIES

Practical Classics on MGB Restoration
Practical Classics on Midget/Sprite Restoration
Practical Classics on Mini Cooper Restoration
Practical Classics on Morris Minor Restoration
Practical Classics on Landrover Restoration
Practical Classics on V W Beele Restoration

BROOKLANDS MILITARY VEHICLES SERIES

Allied Military Vehicles Collection No. 1
Allied Military Vehicles Collection No. 2
Dodge Military Vehicles Collection No. 1
Military Jeeps 1941-1945
Off Road Jeeps 1944-1971
V W Kugelwagen 1940-1975

ACKNOWLEDGEMENTS

The Triumph GT6 made its debut late in 1966 and remained in continuous production until December 1973 by which time more than 40,000 had been manufactured. The majority were exported to the United States.

The car from the beginning was a good looking fastback sports coupé designed for two people only and with adequate power for its size and weight. Its closest counterpart in the early days was the MGB GT which came out a year earlier at the London Motor Show of 1965 and incredibly survived the GT6 by some seven years. Comparison tests from both sides of the Atlantic have been included.

Just as cars evolve due to circumstances, the GT6 became a possibility because of the existence of the Herald/Spitfire and the Triumph 2000, so it is with books. Brooklands Books have evolved over many years fulfilling a small specialised need for information required by second and subsequent owners of collectable cars. The articles that we reproduced here for enthusiasts are all copyright and can only be made available in this form because of the understanding of the original publishers. I am sure that GT6 restorers, owners and would be owners will wish to join with me in thanking the management of the journals listed here for their thoughtfulness and generosity, Autocar, Autosport, Car, Cars & Car Conversions, Modern Motor, Motor, Motor Racing, Motor Sport, Road & Track, Road Test and Wheels. We are indebted to Bill Sunderland and his colleagues, in the Triumph Sports Six Club for their help and support and our thanks also go to Peter Williams for supplying our cover photograph depicting his splendidly preserved 1968 GT6 Mk1.

R.M. Clarke

ACKNOWLEDGEMENTS

The Triumph GT6 made its debut late in 1966 and remained in continuous production until December 1973 by which time more than 40,000 had been manufactured. The majority were exported to the United States.

The car from the beginning was a good looking fastback sports coupé designed for two people only and with adequate power for its size and weight. Its closest counterpart in the early days was the MGB GT which came out a year earlier at the London Motor Show of 1965 and incredibly survived the GT6 by some seven years. Comparison tests from both sides of the Atlantic have been included.

Recently I approached a well known dealer who specialises in sports cars with a view to photographing a good example for the cover of this book. Their reply was an indication of the popularity of the GT6 and can be read as a compliment. 'Unfortunately we do not have one in stock and when we are lucky enough to get one, the chances are that it will not remain on the forecourt for very long'.

Just as cars evolve due to circumstances, the GT6 became a possibility because of the existance of the Herald/Spitfire and the Triumph 2000, so it is with books. Brooklands Books have evolved over many years fulfilling a small specialised need for information required by second and subsequent owners of collectable cars. The articles that are reproduced here for enthusiasts are all copyright and they can only be made available in this form because of the understanding of the original publishers. I am sure that GT6 restorers, owners and would be owners will wish to join with me in thanking the management of the journals listed here for their thoughtfulness and generosity, Autocar, Autosport, Car, Cars & Car Conversions, Modern Motor, Motor, Motor Racing, Motor Sport, Road & Track, Road Test and Wheels. I am indebted to Barry Crawley of Hertfordshire Sports Cars who kindly located the GT6 that is featured on the front cover.

R.M. Clarke

TRIUMPH GT SIX

To the popular Spitfire add a sleek hardtop and the 2-liter 6-cyl engine

BY JOE LOWREY

WHEN TRIUMPH CHRISTENED the little 1147-cc sports 2-seater the Spitfire 4, lots of people jumped to the conclusion that there must eventually be a Spitfire 6 too. After all, the Spitfire had originated as a sort of sawn-off Herald sedan with several vertebrae sliced out of the center of the backbone frame and that same frame also serves for the 1596-cc Vitesse 6 sedan.

Well, it's been a long time coming but that hoped-for 2-seater six has arrived. Not with the Spitfire 6 name which had been expected but christened the Triumph GT Six and equipped with a larger engine than the Vitesse, its body an all-steel version of the fast-back coupe shape which (as fiberglass prototypes) appeared on Spitfires raced at Le Mans in 1964. Instead of the 1596-cc Vitesse engine being warmed up to sports tune, the 1998-cc 90-bhp engine from the Triumph 2000 sedan has been installed in a body trimmed for comfort rather than to save the last ounce of weight.

To me, the most charming kind of sports car has always been the little 'un that really goes, so I had high expectations when I sneaked into a GT Six driving seat back in July with the Triumph Engineering Director, Harry Webster, alongside me. It's a long way down into this little sports car and when you've sat in its shapely bucket seat you then coax your legs in through the door after you. There's a nice leather-covered lightweight steering wheel, carpet on the floor, sun visors above the windshield, an impressive outlook forward over quite a long hood which has a central power bulge to accommodate the 6-cyl engine. There's no unnecessary bulk, though, in a car 145 in. long and 57 in. wide.

Comparisons with the MGB are going to be inevitable and the Triumph puts 11% more displacement (spread over 50% more cylinders but delivering 5 bhp less power in production tune) into a car of 8 in. less wheelbase and overall length, 2.75 in. less width and 1.5 in. less height. There isn't the slightest pretense of a back seat in this Triumph such as the MGB in its closed form offers, just plenty of legroom for two. What the GT Six does have is a top-hinged rear door for luggage, giving access to a flat floor covered with pile carpet: beneath the luggage floor are two parcel compartments, accessible (as also is the luggage area) from inside the car. Altogether, arrangements are reminiscent of the E-type Jaguar.

Driving the GT Six out of an experimental shop congested with other prototypes which I wasn't supposed to look at reminded me of the fabulous steering angles which on a car with only an 83-in. wheelbase produce a turning circle below 25 ft diameter. Out on the Coventry by-pass highway, I realized that this little car is geared to use torque rather than rpm. The stock axle ratio is 3.27 and the car I was driving had the optional overdrive giving a switch-controlled hot shift between ratios of 3.89 or 3.19.

Turning a still-secret car (it is scheduled for U.S. introduction after the first of the year) onto less-public and less-straight roads I found that just switching between 3rd and overdrive-3rd ratios of 4.88 and 4.0 sufficed for a very wide range of car speeds without need to shift gears at all. The car I was driving was just off the assembly line, an early example upon which the engineers were making a quality study, so rather stiff action of the stick shift on a new all-synchromesh 4-speed transmission should not be typical of later, run-in cars.

Around turns, this GT Six on radial-ply tires (which are regular equipment) just seemed to go where I pointed it. I didn't have time to explore its behavior close to adhesion limits but the Triumph engineers (who know the limitations of this model's simple swing-axle independent rear suspension as compared with the semi-trailing link system on their larger and heavier TR-4A) reckon that extra front end weight makes this GT Six behave more gently than a Spitfire 4 when pushed to breakaway. Extra engine length has not involved any wheelbase increase, the radiator moves forward slightly and the firewall is altered also without reducing legroom at each side of the transmission: brake dimensions (discs in front and drums astern) do of course get enlarged.

Because it has such a smooth and quiet 6-cyl engine geared to run at low rpm, the GT Six does not feel as fast as it is: peak power actually comes at rather beyond 100 mph and then there remains another 500 rpm before you reach even the cautionary yellow mark on the tachometer.

Without any doubt, examples of this 2-liter Triumph GT Six are going to be tuned and lightened for competition by some buyers but that isn't the model's primary objective. This is intended to be an enjoyable car to drive which can meet any silencing and atmospheric pollution requirement, one which is comfortable and reliable for everyday transportation of two people—yet really goes.

A 2-litre GT from Triumph

Modified 2000 engine in a Spitfire derivative—

NEW CARS

NEW from Triumph is a two-litre six-cylinder GT model using a high-compression version of the 2000 engine in a chassis and body developed from the Spitfires which ran so well at Le Mans in 1964 and '65. Named the Triumph GT6, its three-figure total price of £985 1s. and unique specification place it in a class of its own. Its maximum speed is given by the makers as 107 m.p.h.

Also notable in the Triumph range for '67 is the latest Vitesse which now has the same high-compression, two-litre version of the 2000 six-cylinder engine as the GT6 instead of the special under-bored 1,596 c.c. edition used to date. In this case, a maximum speed of 100 m.p.h. is claimed. Although the total price of the saloon (at £838 15s. 7d.) is some £67 higher than that of the 1.6-litre model (now superseded), the increase covers improvements to the clutch, gearbox, brakes and body details as well as the bigger engine—and still leaves the Vitesse the cheapest six-cylinder car on the British market by a margin of more than £100. As before, there is a convertible and the total price in this case is £883 0s. 7d.

The GT6 in detail

The new GT6 is a typically Triumph enterprise in the sense that it offers something not found in any other maker's catalogue: a two-seater closed body of the GT type, independent suspension all round and a six-cylinder engine—all at a total price of under £1,000. To obtain a similar combination elsewhere it is necessary to pay a much higher price.

The GT6, in fact, is a sort of "business-man's express" for the less-affluent business-man. Some might think that the 70 m.p.h. limit makes this an odd time to introduce such a car. There are several answers to that. One is that our new antiquated limit does not apply in most overseas markets, for which the first six months' production is in any case being reserved; another is that British people do take their cars abroad (and will go on doing so even on £50 a head plus £25 for the car); and a third point is that the new model should get to the legal limit in Britain very quickly indeed and hold it with a lot in hand.

So far, our actual experience of the new model is confined to a short trial with no opportunities for taking performance figures (see the editor's impressions at the end of this description) but the following figures issued by the makers give a clue to the sort of performance to be expected: top-gear acceleration from 20 to 40 m.p.h., 7.5 sec.; from 30 to 50 m.p.h., 7.4 sec.; from 40 to 60 m.p.h., 7.5 sec.; acceleration through the gears from rest to 50 m.p.h., 7.8 sec.; and to 60 m.p.h., 11.1 sec. The maximum is given as 107 m.p.h.

Thanks to overall dimensions identical with the Spitfire and a kerb weight of 17 cwt., it has been possible to get this sort of performance without recourse to any drastic changes which would spoil the sweetness and flexibility of the 2000 engine. All that has been done is to raise the compression by half a ratio to 9.5:1. The result has been to increase the output from 90 to 95 b.h.p. net at the same speed (5,000 r.p.m.), leaving the maximum torque virtually unchanged. With the relatively high gearing made possible by the power/weight ratio, 1,000 engine r.p.m. is equivalent to 20.15 m.p.h., giving 101 m.p.h. at peak revs., while maximum torque (at 3,000 r.p.m.) occurs at just over 60 m.p.h.

The high-compression GT6 engine is distinguished by a chromium-plated valve cover and drops neatly into a new chassis frame of similar general configuration to that of the Spitfire but modified to cope with the extra size, weight and output of the 2000 unit, still leaving room for an overdrive when required. A slight "power bulge" is necessary in the bonnet top, but this is not so pronounced as to spoil either appearance or forward vision. As on the Spitfire and

Left: *The compact interior is comfortable and some trouble has obviously been taken to make the seats just right. Passenger grab handle is padded.*

Below: *This picture taken through the opening rear window gives a good idea of its size, and of the luggage area available.*

Herald-based ranges, the bonnet is in one with the front wings and swings up on forward hinges to give exceptionally clear access to the engine and front suspension.

In unit with the engine are an 8½-in. diaphragm spring clutch and an all-synchromesh gearbox controlled by a stubby central gear lever. Laycock de Normanville manually-controlled overdrive is optional and, when fitted, operates on the two upper ratios. In this case, a lower top gear is used, the axle ratio being changed from 3.27:1 to 3.89:1, to give an overdrive top ratio of 3.12:1. Although overdrive is also applied to third gear, this is a driver convenience rather than a performance benefit because the overdrive third ratio happens to come out identical with top.

As with the Spitfire, a short open propeller shaft takes the drive to an inboard-mounted hypoid bevel unit (of more robust size), whence swing axles pass on the power to the road wheels via Dunlop SP41 tyres on 4½J rims. Wire wheels are an option.

The suspension arrangements also follow the Spitfire pattern (with appropriate adjustments for weight and so on) but the Girling disc/drum brakes are up in size from Type 12 to Type 16, which have 9.7-in. discs (in place of 9-in.) and 8-in. rear drums in place of 7-in., giving an increase in total rubbed area of just over 30%. The central hand brake, which juts out neatly from the central arm-rest and is conveniently close to the gear lever, has a fly-off ratchet.

The now almost traditional Triumph feature of a steering column adjustable for length (to the extent of 4 in.) and designed to

Quick trip in the GT6

"WE don't mind" said the general manager of Standard-Triumph (a company not noted for its understatements) "comparisons with a miniature E-type or Aston Martin". In a way he was right. So far as one can judge from observation and a 30-mile cross-country journey through the East Midlands and a few laps of the compact Mallory Park circuit, the GT6 bears more than a little resemblance to a scaled down E-type Jaguar coupé. Reducing all dimensions in the proportion of 1.15 to 1 (the ratio of the wheelbases) it should offer comfortable accommodation for two people about 5 ft. 3 in. in height with about 9/10 of the fairly de luxe luggage demanded by a week-ending E-type couple. Passenger capacity in fact exceeds this estimate by a handsome margin, while luggage capacity falls short. A similar comparison of performance based on engine size would predict a maximum speed for the "miniature" (with 0.47 the capacity of the E-type) of 70 m.p.h., acceleration to 50 m.p.h. in 10.2 sec. and a touring fuel consumption of 45.7 m.p.g. The makers' claims, which we have not yet had an opportunity to test, are 107 m.p.h. and 7.8 sec. in respect of maximum and acceleration to 50 m.p.h.

There are more points of similarity in the road behaviour, for the GT6 rides well for a car of its size and type and, on the road at least, has handling to match its performance. The steering is extremely light and precise, cornering is flat and not much perturbed on a dry surface by bumps or camber and there is only a mild tendency for understeer to turn to oversteer if the driver loses his nerve and flannels in the middle of a corner. On a racing circuit the limitations of a simple swing axle, set up for touring use are more evident and "nosing in" is very pronounced—on a wet road it might be embarrassing. We spent most of the time on the 180° south curve at Mallory Park hovering between power-on and power-off in search of a neutral and predictable line. Back on the road this sudden transition was no longer worrying and the new Triumph proved a very fast A to B vehicle, whether one made good use of the engine rev. range, with a maximum 5,500 r.p.m. and very little fuss, or behaved thoroughly idly, taking advantage of the low speed smoothness of a 2-litre six. Only the braking, in which there is no room for a compromise based on scale effect, fell short of the 100 m.p.h. standard on our trial car. Ample stopping power was marred by a tendency to pull unevenly under hard applications of the brakes from high speed.

R.B.-S.

The power bulge on the bonnet top is a genuine one!

A 2-litre GT from Triumph

continued

Space at the back for hiding cameras and various odds and ends.

How it all fits together. Note the spare wheel and fuel tank side by side to give a flat rear floor, the separate chassis members and suspension units. Practically every part of the engine and its ancillary units is instantly accessible for servicing.

telescope on impact, is incorporated in the rack-and-pinion system and an attractive three-spoke wheel with leather-covered rim is used. From lock to lock involves 4¼ turns, which sounds very low-geared for a car of this type—but is not because of the very small turning circle of 25 ft. 3 in. between kerbs; wheel movement for normal turns is more reasonable.

The body is an example of racing experience being put to practical use in a subsequent production model. Styled by Giovanni Michelotti of Turin, the steel-panelled body is a development of the Spitfire but with a new bonnet top and a gently curved roof which slopes down in a single sweep from the top of the curved windscreen to the tail. It has a very large rear window in a 34-in.-wide, top-hinged, counter-balanced rear door giving access to the luggage space behind the seats. This luggage floor is 41½ in. from front to rear and 42 in. wide, with a maximum height above it of 24 in. The result is to give 14.2 cu. ft. of luggage space.

Beneath the rear part of the luggage floor are the tank on one side and the spare wheel, lying horizontally, on the other. Access to the wheel is by a removable panel secured by two screws and a pair of clips, and there is room round the wheel for tools and similar oddments. Below the forward portion of the luggage floor and accessible from behind the seat squabs is a useful parcel shelf which supplements the front parcel shelves under the facia board.

Apart from a slight obstruction caused by the rear quarter panels, vision is very good but a user who regularly carried much luggage would probably need to fit an external mirror. The door windows are of the winding type and a good point is that the rear quarter lights are pivoted for

The lift-up front cowling gives splendid access to the six-cylinder engine which fits snugly in place with only a slight bulge on the bonnet top to give away its presence.

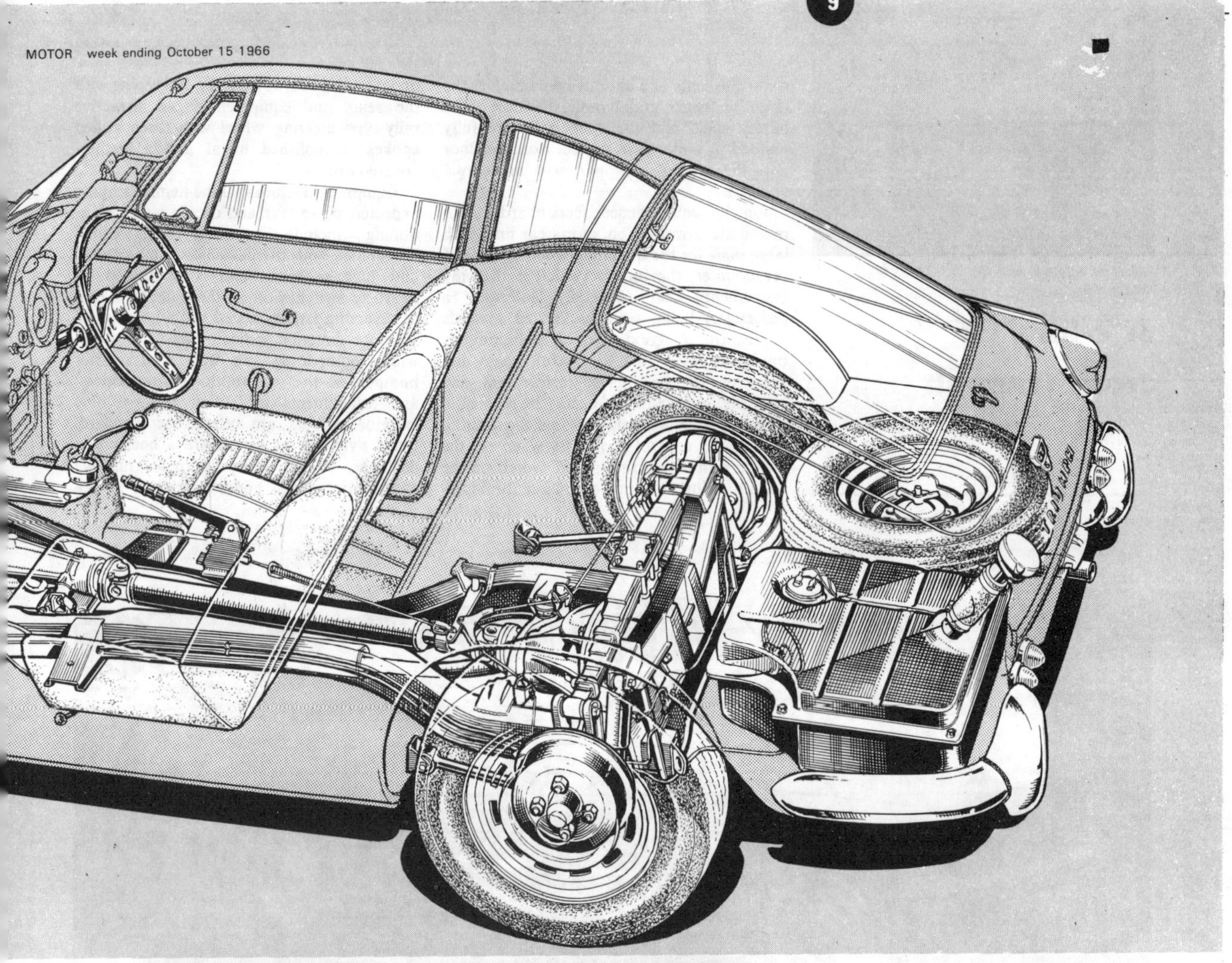

The rakish hardtop has a really large rear window and the sides have a decidedly aircraft look about them.

ventilation as well as the hinged panels on the leading edges of the doors.

Inside, the body is compact but not cramped. Doors 29 in. wide at waist level, with burst-proof latches, give access to an interior 45 in. wide between doors and 38 in. high, of which 34 in. represents headroom above the seat cushion. The seats themselves are of the rally type and measure 20 in. fore and aft and 19½ in. in width. The usual fore-and-aft adjustment is provided and both squabs tilt forwards for access to the luggage space. Ambla material is used to

Specification

Engine

Cylinders	6 in-line with 4-bearing crankshaft
Bore and stroke	74.7 mm. x 76 mm. (2.94 in. x 2.99 in.)
Cubic capacity	1,998 c.c. (122 cu. in)
Piston area	40.7 sq. in.
Compression ratio	9.5:1
Valvegear	In-line o.h.v. operated by pushrods and rockers
Carburation	Two Stromberg 1.50 CD sidedraught carburetters, fed by mechanical pump, from 9¾-gallon tank
Ignition	12-volt coil, centrifugal and vacuum timing control, 14 mm. Champion sparking plugs
Lubrication	Eccentric lobe pump, AC full-flow filter and 11-pint sump
Cooling	Water cooling with pump, fan and thermostat; 11-pint water capacity
Electrical system	12-volt 48 amp. hr. battery charged by 300-watt generator
Maximum power	95 b.h.p. net at 5,000 r.p.m. equivalent to 124 lb./sq. in. b.m.e.p. at 2,490 ft./min. piston speed and 2.33 b.h.p. per sq. in. of piston area.
Maximum torque	117.3 lb. ft. at 3,000 r.p.m., equivalent to 145 lb./sq. in. b.m.e.p. at 1,490 ft./min. piston speed

Transmission

Clutch	8½-in. diaphragm type
Gearbox	4-speed with synchromesh on all forward gears
Overall ratios	3.27, 4.11, 5.82 and 8.66; reverse, 10.15. (With optional overdrive, 3.89 (O/D 3.12), 4.86 (O/D, 3.89), 6.92 and 10.30; reverse, 12.06)
Propeller shaft	BRD open
Final drive	Inboard-mounted hypoid bevel and swing axles

Chassis

Brakes	Girling hydraulic, disc-front/drum-rear
Brake dimensions	Front 9.7 in. dia. discs; rear 8 in. dia. x 1¼ in. wide drums
Brake areas	60.2 sq. in. of lining (22.2 sq. in. front plus 38 sq. in. rear) working on 260 sq. in. rubbed area of discs and drums
Front suspension	Independent, by coil springs and wishbones with Girling telescopic dampers and anti-roll bar
Rear suspension	Independent by transverse leaf springs, radius arms and swing axles. Girling telescopic dampers
Wheels and tyres	Steel disc wheels with 4½J rims and 155-13 SP 41 tyres
Steering	Alford and Alder rack-and-pinion with adjustable collapsible column

Dimensions

Length	Overall 12 ft. 1 in; wheelbase 6 ft. 11 in.
Width	Overall 4 ft. 9 in; track 4 ft. 1 in. at front and 4 ft. 0 in. at rear
Height	3 ft. 11 in; ground clearance 4 in. laden
Turning circle	25¼ ft.
Kerb weight	17 cwt. (with fuel, oil, water, tools, spare wheel, etc.)

Effective gearing

Top gear ratio	20.15 m.p.h. at 1,000 r.p.m. and 40.2 m.p.h. at 1,000 ft./min. piston speed; with optional overdrive, 16.8 m.p.h. (O/D 21.0 m.p.h.) and 33.6 m.p.h. (O/D, 42.1 m.p.h.)

Tail end treatment, with the GT6 badge in the middle.

A 2-litre GT from Triumph

continued

cover the seats and various items such as the large passenger grab handle, the front of the parcel shelf and the waist rails are fully padded in material to match. Both the floor and the luggage compartment are carpet covered.

On the walnut veneer facia board, instruments are arranged on sports car lines with large dials for the matching speedometer and rev counter, the former (which is the only instrument that normally interests a passenger) being sensibly located towards the centre. A good deal of attention has obviously been paid to careful design and grouping of the smaller instruments and controls, an example being the use of a finger-tip side/head/dip lever on the steering column to control the lights when on the move, with a separate on/off switch on the facia to bring them into operation; the finger-tip lever also acts as a flasher. In keeping with the seats and equipment is an attractive rally-type steering wheel with three drilled spokes in polished metal and a leather-covered rim.

Equipment includes all the items normally expected nowadays and details worth mentioning include a roof light which illuminates both the passenger compartment and luggage space and is controlled by courtesy switches on all three doors, a pair of reversing lamps, sealed beam headlamps, separate direction indicator flashers, a full-width bumper at the front and quarter bumpers at the rear (both with overriders) and attachments for safety harness. Five exterior colours are offered as standard. Items available as extras are a heater and demister, radio, safety harness, luggage straps, wire wheels and overdrive. M

Outwardly it could be just another Vitesse but the bonnet houses a lot more power.

The Vitesse

EVER since its introduction four and a half years ago, the Vitesse has provoked very divided opinions. Some drivers have realized that the maker's object was to produce a sort of super Herald with a better, but above all a more-refined, performance. Others have mistakenly expected GT performance and been disappointed. The new Vitesse 2-litre should satisfy both schools of thought because the refinement and performance are both there in greater measure. The maximum speed is claimed to be 100 m.p.h. and acceleration should not be so far short of that of the new GT6 because the 2-litre, six-cylinder engine is in the same stage of of tune and the difference is accounted for only by some extra frontal area, a less aerodynamic shape and a kerb weight of $18\frac{1}{4}$ cwt. compared with the GT's 17 cwt.

Externally, the new Vitesse looks just like the old apart from the addition of a reversing light and new emblems front and rear. Inside the body, comfort has been improved by the fitting of new seats in both front and rear compartments while a three-spoke leather-trimmed steering wheel is used. A rev counter is standard equipment.

"Invisible" improvements include changes in the transmission and brakes. To cope with the extra performance, the clutch has been increased in diameter by $\frac{1}{2}$ in. to $8\frac{1}{2}$ in. and is now of the diaphragm type. In unit with it is an improved gearbox which now has synchromesh on all forward gears. As before, it can be supplemented (as an extra) by a Laycock de Normanville overdrive operating on the two upper ratios. A more substantial final-drive unit and larger-diameter half-shafts are also used.

The dimensions of the disc-front/drum-rear brakes have likewise been increased to cope with the added performance and are now of the same size as those of the new GT, with 9.7 in. discs at the front and 8 in. x $1\frac{1}{4}$ in. drums at the rear.

As before, both saloon and convertible models are offered and buyers have the choice of 10 colour schemes. Extras offered include safety harness, a sunshine roof for the saloon, Armstrong GT7 heavy-duty dampers, heavy-duty tyres, a special thermostat for winter conditions and a sump guard. M

Brief Specification

Engine:

6-cyl., 74.7 mm. x 76 mm., 1,998 c.c., o.h. valves (push rods); two side-draught Stromberg 1.50 CD carburetters; 95 b.h.p. net at 5,000 r.p.m.; 117.3 lb. ft. torque at 3,000 r.p.m.

Transmission:

$8\frac{1}{2}$-in. diaphragm-type clutch; 4-speed gearbox with synchromesh on all forward gears; ratios, 3.89, 4.86, 6.92 and 10.31; reverse, 12.06. Laycock de Normanville overdrive available as optional extra giving O/D top of 3.11 and O/D 3rd of 3.89. Road speed at 1,000 r.p.m., 17.3 m.p.h. (O/D, 21.6 m.p.h.)

Running gear:

Girling disc-front/drum-rear brakes; independent coil-and-wishbone front suspension; independent rear suspension by transverse leaf spring, swing axles and radius rods; rack-and-pinion steering. 5.60-13 tubeless tyres.

Dimensions:

Length, 12 ft. 9 in.; Width, 5 ft. 0 in.; Kerb weight, $18\frac{1}{4}$ cwt.

SPITFIRE +

Triumph's new GT6 has a character all its own

PERSEVERING DOGGEDLY with our own personal Motor Show so as to catch the deadline for this issue, we turned up at STI in Coventry one bleak late-summer day in search of inspiration. A sour-faced commissionaire glowered at us from his varnished pulpit in the gaol-like reception parlour and, after informing us abruptly that we were late, proceeded to spin an elaborate yarn to the effect that the press people we had come to see had all gone to lunch. After keeping us thus imprisoned for 20 minutes he finally directed us to a cheerful new building by the sports ground where house-magazine editor Arnold Bolton had been fuming all the while. How in hell does this god-forsaken country survive when it is dominated by pumped-up Hitlers like that commissionaire? 'If we sacked him for it they'd all be out' said Arnold bitterly. We can imagine.

Luckily for everybody concerned the top brass at STI don't allow themselves to be deterred by such trivia, and what we found when we finally got the mutual recriminations over was a brace of pretty sophisticated sporting '67s. Standard Triumph was among the first manufacturers to realise the benefits of standardisation without loss of character (remember the Vanguard-powered Renown?), and today the production boffins there are real experts at getting their money's worth from every single component – which is why you find Herald front suspensions, steering racks, minor controls, switchgear and so forth in almost every kitcar in the country. It all pays the bills.

Indeed the remarkable thing about the latest GT6 and Vitesse-2 is that they use so very few brand new mechanical components. What the designers have done is to shuffle their existing catalogue of bits and pieces to produce combinations which appeal to a different market not only because they look different but because they *are* different. This is the Detroit rather than the Longbridge approach, and it will pay off – without introducing any unwelcome Americanisation of the product.

Looking at the Vitesse first because it is simpler, we find the two-litre, six-cylinder engine from the current 2000 saloon occupying a space but newly vacated by its smaller but similar sister. Wasn't the Vitesse powerful enough as it was? On paper, yes. But CAR in company with a lot of customers found that the original model never really fulfilled its potential from a performance point of view, and anyway there was occasional →

→ resistance to the idea of a small and apparently hard working Little Six because of the memories it evoked of certain less successful products of the late 1930s. The old Vitesse felt constricted when it should have felt free, hard-working when it should have seemed relaxed. As a two-litre it naturally inherits the effortless feel of the current 2000 as well as the ability to ease its small and very light Herald-based bodyshell through the air at an appropriately brisker clip than it could ever reasonably attain in four-cylinder form even with heavy help from the go-faster fraternity. And on top of all this it allows the company to rationalise with a single six-cylinder engine.

Chassis modifications are unnecessary, strictly speaking, since the bigger engine weighs no more than its predecessor. But we were delighted to find that STI had found it worthwhile to try and improve the Vitesse's tail-happy handling. Using Goodyear G800 tyres the saloon now feels fully able to cope with the attentions of a skilled driver, at least in the dry, although the convertible, with its less favourable weight distribution and less rigid structure, feels distinctly ill at ease. Ride is little better than before, but an entirely new gearbox (shared with the GT-6) contributes to the overall improvement by providing synchromesh on bottom gear. Overdrive is available, but the one on our car objected to strenuous use and packed up after a few minutes. Steering remains accurate but heavy, and we noticed none of the dreaded rack rattle.

Inside, the Vitesse boasts new seats which are actually better in shape (less of that overstuffed feeling, and more location from the backrests) but no more satisfactory in terms of attitude. You still sit too upright in this car, and there is still not enough rearward adjustment. (It's easy to see why when you venture into the back, where legroom really is at a premium.) Sad, too, that the once-vaunted 'infinitely variable' front seat adjustment mechanism has become refined to the point where it merely gives an extremely limited range of decreasingly practicable angles. But the trim is still neat and tidy, there's polished wood for those who fancy their chances of avoiding violent contact in a crash, and a rev counter is standard in this model as it has been for the past couple of seasons. All of which indicates that the Vitesse will still have some appeal as one of the very few compact sporting saloons on the market. And the new engine coupled with that famous 25ft lock ought to make it just about the quickest London traffic-car under £1000.

That is, if you ignore its sparkling sister the GT-6. This one has no family pretensions (the lack of even a token jump-seat is among the first things you notice, especially as it looks as if there was supposed to be one at some point) but as a strict two-seater will repay fairly close study. The engine differs from the one in the Vitesse only in the colour of its rocker-box, and the chassis is more or less Spitfire apart from spring rates and rear suspension mountings. These are roughly what we think they ought to be in all the other models as well, with a bit of built-in negative camber plus fat tyres and more sympathetic ratings all round. But can they cope with two litres and 95bhp in a 1900lb car?

A short run round the Warwickshire lanes enlightened us somewhat on this score. In fact we were startled to find how *entirely* different the GT-6 feels from the Spitfire on which it is based – a much greater difference, for example, than between a Herald and a Vitesse. The view outwards is altered immeasurably because of the different cockpit and somewhat beetle-browed fastback roofline, the pudgy new seats (again a shade too upright, and non-adjustable for rake) give a big-car feel, steering is heavier thanks to fatter tyres and much greater →

This page, top: Triumph's biggest passenger-car engine slots neatly into the Spitfire's generous underbonnet area to create Britain's first hotrod

Above: With the barely distinguishable two-litre Vitesse in the background, GT-6 shows off basically beautiful lines marred by detail clutter

Right: the rev counter's been there for some time, but Formula leather-rim is a production first and Vitesse seat frames have new padding for '67

Left: Low, lithe look befits a challenger to BMC's beautiful MGB GT, although extra metal in roof pressing creates an acute visibility problem for tall men

Left, below: would you recognise it in your mirror? Full-width slatted grille and bonnet-top bulge are GT-6 giveaways, but we had hoped for more identification

Below, right: easy access to the biggest baggage area in any small sports car, with perhaps enough room for a dog or even the odd inebriated party guest

Bottom, right: plenty to look at and play with in GT-6 office, including one of those wheels. Visibility is restricted, door handles well hidden

photographed by Charles Pocklington at The Bear Inn, Berkswell, Warwickshire

front-end weight, and in general the whole car feels solider and more substantial. This is perhaps a good thing, since the transplanted power plant certainly pours out some poke. Instinct suggests acceleration of about the TR4 order with a top speed in overdrive (an option which we discovered too late to use properly!) well over the ton.

As you would expect, the engine feels as if it is working well within limits most of the time. This is more than one can say for the chassis, which we managed to provoke into a couple of fair old tail-slides until a loud knocking noise from behind the seats (wheel or driveshaft rubbing on something?) discouraged us from trying harder. Not that this is likely to persist in the production cars (ours was a series prototype) and not that we are trying to suggest the GT-6 is overpowered, but it will certainly be a car to treat with respect.

As for its looks, everyone here is unanimous in declaring the tin roof a Good Thing from an aesthetic as well as an aerodynamic point of view. But we think buyers in the near-£1000 class are entitled to expect better detail design, with less of a clutter of lights and badges on that shapely rump, as well as perhaps more model identification in front to distinguish it in the mirrors of those who are about to be overtaken from the slower and much cheaper Spitfire. Wire wheels would have been a nice gesture at the price.

But who's for a Giant Test between this and the MGB?

Road Test by JOHN BOLSTER

THE TRIUMPH GT6

IT is splendid fun to have a small engine, tuned as for racing, which only gives its designed performance if the gear lever is used continuously. Fun, that is, for week-end jaunts, but a car so powered can be a tiring companion for everyday use, wearing itself out quickly and wearing its driver out, too.

In the early days of motoring, high-performance cars had light bodies and big engines, doing their work easily on a high gear. In the intervening period, a sports car was not regarded as being worthy of note unless its engine turned at many thousands of revolutions per minute. Suddenly, there has been a return to sanity, for the ancients were right and a good big 'un will always beat a good little 'un.

Yet the ideal car for 1967 must have compact overall dimensions, a big machine being tiresome in modern traffic conditions. The Triumph solution is to produce a car no bigger than the popular 1100 cc or 1200 cc sports cars but fitted with a 2-litre, 6-cylinder engine. It is those six cylinders which give the car its special character. Driving hard with full use of the gears, one achieves the performance which one would expect of a well-streamlined 2-litre sports car weighing only 17 cwt. However, by remaining in the very high top gear, the driver still obtains almost as much performance with remarkable ease, silence and economy.

The separate wasp-waisted Spitfire chassis frame is used, only the engine mounting brackets being different. The radiator for the closed-circuit cooling system is carried much farther forward, and a closed-circuit breather system, with a Smith valve, ensures that engine smells will not seep through from the very crowded bonnet space. The Spitfire type of bonnet has been retained, giving superb accessibility to the engine and front suspension assembly.

A new close-ratio gearbox, with synchromesh on all four gears, has been designed for the GT6. It is mated with the almost "square" 1998 cc six-cylinder engine, which has a 9.5 to 1 compression ratio for this application and develops a net power output of 95 bhp at 5000 rpm. A new heavy-duty chassis-mounted hypoid unit has a 3.27 to 1 ratio, or 3.89 to 1 with the 0.80 to 1 overdrive. The suspension, independent all round, is by wishbones and helical springs with an anti-roll bar in front, the rear transverse spring supporting swing axles.

The fixed-head two-seater coupé body has been developed from the Le Mans Spitfires. It has an opening rear panel containing the sloping rear window, which exposes a very large luggage platform. Unfortunately, the luggage or parcels can be seen through the window and some sort of cover might be arranged to conceal these from miscreants, who think nothing of smashing a window nowadays, even in well-lit London streets.

The body is by no means cramped, though the low roof gives a rather beetle-browed effect. I am 6 ft tall, with rather wide shoulders and size 10 shoes, but I was perfectly comfortable for long journeys. Curiously enough, I have insufficient room in the Spitfire, but I believe the steering wheel—attractively leather-covered with spokes that do not obscure the instruments —is smaller in the case of the GT6 and perhaps the seating has been altered for the better.

Comfortably placed with an adequate all-round view, the driver can settle down to enjoy himself. Though bottom gear is exceptionally high, it is quite easy to spin the wheels, even on dry roads. The gear changes go through quickly and easily, but one tends to change up early because the

top gear performance is so good. The test car had the optional overdrive but the GT6 may be ordered without this component, when the higher geared (lower numerical ratio) hypoid unit gives substantially the same engine revolutions at maximum speed. With so much engine torque, the extra £58 7*s* 8*d* is perhaps hardly justified.

Very smooth throughout its range, the engine has sufficient power to propel the car at a speed just short of 110 mph. The test car had an appreciable propeller shaft period at 95 mph or so, but one would imagine that this fault applied only to an individual vehicle.

The ride is firm, and very little roll is apparent. The suspension is only noticeably hard on poor road surfaces and there is a welcome absence of pitching. Though the GT6 is naturally heavier in front than the Spitfire, it is a well-balanced car, and the tail may be hung out under suitable circumstances. Tested on a racing circuit in wet and slippery conditions, the car gave plenty of confidence and returned good lap times.

This stability was also maintained during hard braking. The Girling brakes, with discs in front, have no servo, but the pedal pressure is by no means heavy and the resistance to fade is marked. Though the rack and pinion steering is perhaps not exceptionally quick, it gives plenty of "feel" and is never twitchy, the car being very easy to hold straight at speed, even on highly cambered roads or in gusty side winds. The traditional fantastic steering lock of the smaller Triumphs is retained, a feature that I blessed every time I parked the car in London.

Considering the low roof-line, this is not a difficult car to enter and leave, while it is very easy to stow heavy luggage. An Italian designer might have managed a little more glass area for the windscreen, but the rear view mirror gives a splendid panorama through the sloping window which, as is usual with this type of body, keeps remarkably clean. The heater—quoted as an extra, curiously enough—makes the interior very cosy, but better ventilation and demisting would not come amiss.

Some people have expressed a desire for a soft-top version. I am afraid that open cars are gradually on the way out, both because the insurance companies dislike them and as a result of their rapid deterioration. There is also the fear that roll-over bars may be made compulsory, and nothing looks more unsightly than an open car so fitted. I think that the makers have done well to construct this body in permanently closed form, therefore.

From the point of view of appearance, the GT6 is a definite success, the difficult rear end problem having been solved satisfactorily. The test car was very well finished in an attractive dark blue, and my only criticism of its looks concerns the bright strips along the tops of the wings, which are perhaps a bit prominent. However, I believe that these are used to secure the panels, rendering quick repairs easy. The body is certainly very well made, with a complete absence of rattles.

The high gearing and flexible top gear performance pay dividends in fuel economy, an easy 30 mpg being possible. As the petrol tank holds 9¾ gallons, the range is satisfactory. Continuous town work does not cause the engine to warm up or misfire, the GT6 proving to be an admirable shopping car.

Whether you regard the new Triumph as a sports car or as a practical two-seater saloon, it proves to be a most attractive machine. Good-looking and a delight to drive, it gains greatly from having a flexible six-cylinder engine. As a compact but very potent high-performance car, the GT6 gives a new experience in motoring, and every enthusiast will want to try it.

SPECIFICATION AND PERFORMANCE DATA

Car tested: Triumph GT6 fixed-head two-seater coupé. Price, with overdrive and heater, £1056 19*s*. 1*d*.

Engine: Six cylinders, 74.7 mm × 76 mm (1998 cc). Pushrod-operated overhead valves. Compression ratio 9.5 to 1. 95 bhp (net) at 5000 rpm. Twin Stromberg horizontal carburetters. Lucas coil and distributor.

Transmission: Borg and Beck diaphragm spring clutch. Four-speed all-synchromesh gearbox with short central lever, fitted Laycock-de Normanville overdrive (extra), ratios 0.8 (overdrive), 1.0, 1.01 (overdrive 3rd), 1.25, 1.78 and 2.65 to 1. Open propeller shaft. Hypoid bevel final drive, ratio 3.89 to 1.

Chassis: Backbone-type chassis with welded steel body. Independent front suspension with wishbones, helical springs and anti-roll bar. Rack and pinion steering. Independent rear suspension with swing axles and transverse leaf spring. Telescopic dampers all round. 9.7 ins diameter front disc brakes with 8 ins × 1.25 ins drums at rear. Bolt-on disc wheels, fitted with 155 × 13 ins Dunlop SP41 tyres.

Equipment: 12-volt lighting and starting. Speedometer, rev counter, water temperature and fuel gauges. Heating and demisting (extra). Two-speed windscreen wipers and washers. Flashing direction indicators.

Dimensions: Wheelbase 6 ft 11 ins. Track (front) 4 ft 1½ ins; (rear) 4 ft 0¼ in. Overall length 12 ft 1¼ ins. Width 4 ft 9¼ ins. Turning circle 23 ft. Weight 17 cwt.

Performance: Maximum speed, 108 mph. Speeds in gears: direct top, 100 mph; third, 80 mph; second, 55 mph; first, 34 mph. Standing quarter-mile, 17.9 secs. Acceleration: 0-30 mph, 3.5 secs; 0-50 mph, 7.2 secs; 0-60 mph, 10.4 secs; 0-80 mph, 18.6 secs.

Fuel consumption: 25 to 33 mpg.

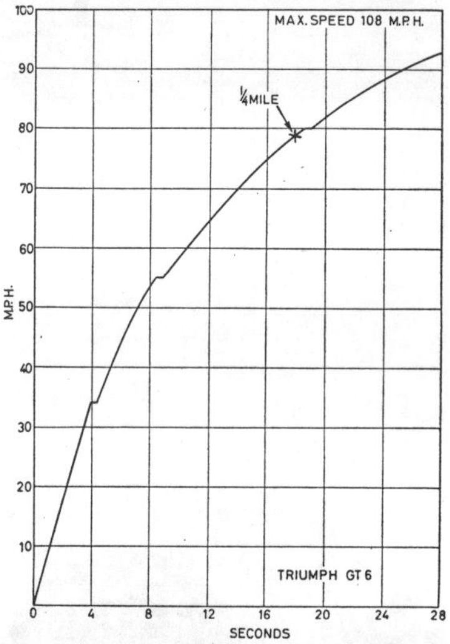

GORDON CHITTENDEN PHOTOS

TRIUMPH GT6

2-liter 6 engine and a hardtop on the Spitfire add up to something distinctive among today's sports cars

IS IT POSSIBLE to take an existing model roadster, graft on a coupe top, insert the engine from a sedan and have the result add up to an entirely new combination in the medium-priced sports/GT car genre? Yes, it is possible and we're happy to report that Triumph has accomplished it with the new GT6, the only car in its class with a "six."

The GT6 derives from the popular Spitfire 4 and shares the same backbone frame and separate lower body with that model. The new hardtop roof is grafted on in the same way as several other current cars and is no more/less successful/

unsuccessful than most. Under the skin there's little change from the Spitfire chassis except for the addition of the undersquare 2-liter 6-cyl engine from the Triumph 2000 sedan. The unequal-arm front, and swing-axle rear, suspension layouts are intact except for such details as spring rates.

We liked the 2-liter six in the 2000 and we like it even better in the GT. Though not a new design, this straightforward pushrod overhead valve unit is one of the quietest and smoothest sixes to be found anywhere and idles at an even 500 rpm with just a trace of clicking from its mechanical tappets. Besides refinement, torque is its stock in trade. It's one of those engines that doesn't seem to care what gear is engaged behind it. At the other end of the scale, it revs freely to 6000 rpm, the top of the orange sector on the tachometer face, and is said to be safe for a sustained 5500.

It's nice to have an engine with enough torque that you don't have to operate the gearbox constantly. When gearchanges are called for, though, the shifting is pleasant, the 8.5-in. diaphragm-spring clutch takes up smoothly, the indirect gears are almost dead silent and the linkage is crisp with fairly short but awkward throws.

Our test car was equipped with the optional ($175 extra) Laycock-de Normanville overdrive which is controlled by a stalk on the right side of the steering column. Engagement takes place about 1.5 sec after the lever is moved and is quite abrupt. There is no inhibitor switch to prevent a downshift to direct drive on the overrun but it is easier on the gears if the driver keeps his foot down at least a little when making a change.

This overdrive isn't essential to satisfactory operation of the GT6, happily. The normal axle ratio is 3.27:1, giving 20.1 mph/1000 rpm, quite acceptable gearing for overall use. In the overdrive-equipped car, a 3.89:1 final drive ratio is fitted and this, combined with the 0.802:1 ratio of the overdrive gear, results in a 3.12:1 overall ratio or 21.3 mph/1000 rpm. Because the gearbox ratios aren't altered, each of the four ratios becomes "tighter" with the overdrive. Our test car could be started easily in 2nd gear, as a matter of fact.

For the U.S., the handsome wire wheels are standard as are the 155-13 Dunlop SP-41 radial ply tires.

We approach any car with conventional swing axles with a little apprehension but we found that the GT6 could not be faulted on its handling. The 6-cyl engine brings front-end weight distribution to 56%, which is probably a good thing with swing axles. For ordinary-to-brisk driving, the car steers neutrally and simply goes where it's steered with great apparent stability. The tail can be brought out at will either by poking the throttle (in the right gear, naturally) or by just tweaking the wheel a little too much. Breakaway is smooth and one gets the feeling that the car has a degree of oversteer that can be enjoyed and utilized by a moderately skilled driver while never crossing up an unskilled one. The GT6 corners flat, too, and doesn't seem to want to lift its inside rear wheel in violent low-speed maneuvers.

Over rough surfaces, the coupe performs creditably too. Its body is tight and the wheels stay on the ground. There is some chatter from the rear suspension but harshness isn't excessive on any surface.

Steady high speed driving is another forte of the GT6. Its overall noise level is low enough to be untiring and the engine isn't working itself to death to maintain a decent cruising speed.

The interior finish also gets high marks. The seats aren't adjustable for rake but are reasonably well formed and pleasing to the eye. The dashboard (fascia, if you insist) is polished walnut, the vinyl upholstery is tastefully patterned and the floor areas are covered with carpeting.

The layout of the instruments isn't as good as it might be. The speedometer and tachometer are placed on the flat panel with little regard for ease of reading and would benefit from a little angling upward and toward each other. Switches are all clearly labeled in sign language and grouped in the center dash section. Water temperature and fuel gauges are also in

TRIUMPH GT6
AT A GLANCE

Price as tested.............................$3309
Engine...............inline 6, ohv, 1998 cc, 95 bhp
Curb weight, lb.................................1970
Top speed, mph.................................107
Acceleration, 0–¼ mi, sec....................18.8
Average fuel consumption, mpg..................24
Summary: Something quite new and distinctive. Lively, quiet and nimble, with an outstandingly smooth engine. Not for tall people.

TRIUMPH GT6

the center section and are legible enough. Oil pressure and generator warning lights are in the lower part of the speedometer face.

With overdrive the steering column is full of branches and twigs. Even after several days of driving we still found ourselves shifting into overdrive when all we wanted to do was dim the headlights. A headlight flasher is incorporated into the light stick, a device we like though it is not strictly legal in all states.

The GT6 isn't a tall man's car, as the Comfort Rating Index shows. Our tallest driver, who is 6 ft 3, found himself too cramped to drive it with any degree of pleasure and our two 6-footers complained of their heads touching the roof. No one liked the location of the accelerator pedal; it was simply too high, and required the leg to be held in an awkward and tiring position. Seems a pity to build such a nice car and then have it be comfortable for only a limited range of drivers.

There is no nonsense about rear seats in the GT6. The large carpeted area behind the seats will take plenty of luggage and there is in addition a cubbyhole that would be handy for storing such things as cameras out of sight.

In keeping with the generally high level of mechanical refinement of the GT6 is the brake system which turned in a creditable 28 ft/sec/sec (87% g) maximum deceleration. In our simulated panic stop from 80 mph, directional control was undisturbed down to about 40, when the rear wheels began to lock. Pedal efforts aren't particularly low (no vacuum booster), but not bothersome either, and the brakes faded only 20% during our 6-stops-from-60 test.

As the whole front end sheet metal hinges forward, service access is of the highest order for the engine and front suspension. A nice touch of elegance under the hood comes from the chromed valve cover. The engine has a sealed (no loss) cooling system and chassis lubrication/oil change intervals are in keeping with the contemporary trends toward ever less fuss over routine maintenance.

The detail styling of the GT6 hasn't been carried out with quite the finesse evident in its powertrain and running gear. The Michelotti-styled lower section is generally pleasant but what detracts from the overall effect is the clumsy details such as multiple light units, chrome-trimmed seams and extraneous lines in curious places.

In summary, the GT6 is a smaller package that incorporates many of the same qualities that make the Jaguar E-type such an exhilarating car. It is smooth; it has good torque, low noise level and agility as well as stability in its handling. It's a great improvement over the Spitfire 4 from which it descended. Not that the Spitfire 4 was bad, it's just that the GT6 is so much better. It has no parallel and it's worth the money.

ROAD TEST

TRIUMPH GT6

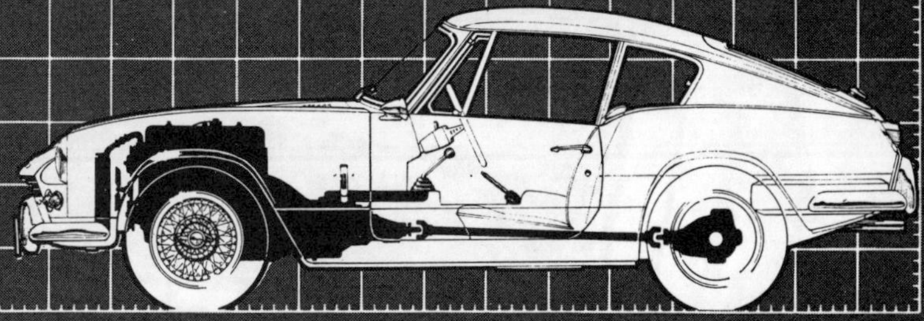

PRICE

Basic list $3039
As tested $3309

ENGINE

Type . . . . . . 6 cyl inline, ohv
Bore x stroke, mm . . . . . 74.7 x 76.0
Equivalent in . . . . . 2.94 x 2.99
Displacement, cc/cu in . . . 1998/122
Compression ratio 9.5:1
Bhp @ rpm . . . . . 95 @ 5000
Equivalent mph 105
Torque @ rpm, lb-ft . . . 117 @ 3000
Equivalent mph 64
Carburetion . . 2 Stromberg 1.50 CD
Type fuel required . . . . . premium

DRIVE TRAIN

Clutch diameter, in 8.5
Gear ratios: o'drive (0.802) . . 3.12:1
4th (1.00) 3.89:1
3rd (1.25) 4.86:1
2nd (1.78) 6.92:1
1st (2.65) 10.3:1
Synchromesh . . . . . on all 4
Final drive ratio 3.89:1
Optional ratio 3.27:1

CHASSIS & BODY

Body/frame: steel backbone frame with separate steel body.
Brake type: front—single caliper discs, 9.7-in. diameter; rear—8.0 x 1.25-in. drums.
Swept area, sq in 260
Wheel type & size, in . . . . . wire, 13x4.5J
Tires . . . . . Dunlop SP-41 155-13
Steering type . . . . . rack & pinion
Overall ratio . . . . . n.a.
Turns, lock-to-lock 4.5
Turning circle, ft 25.2
Front suspension: unequal-length A-arms, coil springs, tube shocks, anti-roll bar.
Rear suspension: swing axles, trailing arms, transverse multi-leaf spring, tube shocks.

OPTIONAL EQUIPMENT

Included in "as tested" price: overdrive, AM radio, outside mirror, 2 lap belts.
Other: luggage straps.

ACCOMMODATION

Seating capacity, persons 2
Seat width . . . . . 2 x 15.5
Head room 38.0
Seat back adjustment, deg . . . none
Driver comfort rating (scale of 100):
Driver 69 in. tall 75
Driver 72 in. tall 70
Driver 75 in. tall 50

INSTRUMENTATION

Instruments: 140-mph speedometer, 7000-rpm tachometer, odometer with trip meter, water temperature, fuel.
Warning lights: directional signals, high beams, generator, oil pressure.

MAINTENANCE

Crankcase capacity, qt 4.8
Change interval, mi 6000
Filter change interval, mi . . . 12,000
Chassis lube interval, mi 6000
Tire pressures, psi 20/24

MISCELLANEOUS

Body styles available: coupe only.
Warranty period, mo/mi: 6/6000 parts & labor; 12/12,000 parts.

GENERAL

Curb weight, lb 1970
Test weight 2290
Weight distribution (with driver), front/rear, % 56/44
Wheelbase, in 83.0
Track, front/rear 49.0/48.0
Overall length 145.0
Width 57.0
Height 47.0
Frontal area, sq ft 14.9
Ground clearance, in 4.0
Overhang, front/rear 28.4/33.6
Usable trunk space, cu ft 6.6
Fuel tank capacity, gal 11.7

CALCULATED DATA

Lb/hp (test wt) 24.1
Mph/1000 rpm (o'drive) 21.3
Engine revs/mi (60 mph) 2810
Piston travel, ft/mi 1400
Rpm @ 2500 ft/min 5020
Equivalent mph 106
Cu ft/ton mi 86.8
R&T wear index 39.4
Brake swept area sq in/ton 227

ROAD TEST RESULTS

ACCELERATION

Time to distance, sec:
0–100 ft 4.0
0–250 ft 6.5
0–500 ft 10.0
0–750 ft 13.1
0–1000 ft 15.8
0–1320 ft (¼ mi) 18.8
Speed at end of ¼ mi, mph 75
Time to speed, sec:
0–30 mph 4.4
0–40 mph 6.5
0–50 mph 8.9
0–60 mph 12.3
0–70 mph 16.4
0–80 mph 21.9
0–100 mph 31.9
Passing exposure time, sec:
To pass car going 50 mph 7.5

FUEL CONSUMPTION

Normal driving, mpg 22–26
Cruising range, mi 250–300

SPEEDS IN GEARS

O'drive (5080 rpm), mph 107
4th (6000) 103
3rd (6000) 82
2nd (6000) 57
1st (6000) 38

BRAKES

Panic stop from 80 mph:
Deceleration, % g 87
Control . . . . . good
Fade test: percent of increase in pedal effort required to maintain 50%-g deceleration rate in six stops from 60 mph 20
Parking brake: hold 30% grade . . . . . yes
Overall brake rating . . . . . very good

SPEEDOMETER ERROR

30 mph indicated . . . . . actual 28.6
40 mph 38.6
60 mph 58.6
80 mph 78.3
100 mph 98.1

ACCELERATION & COASTING

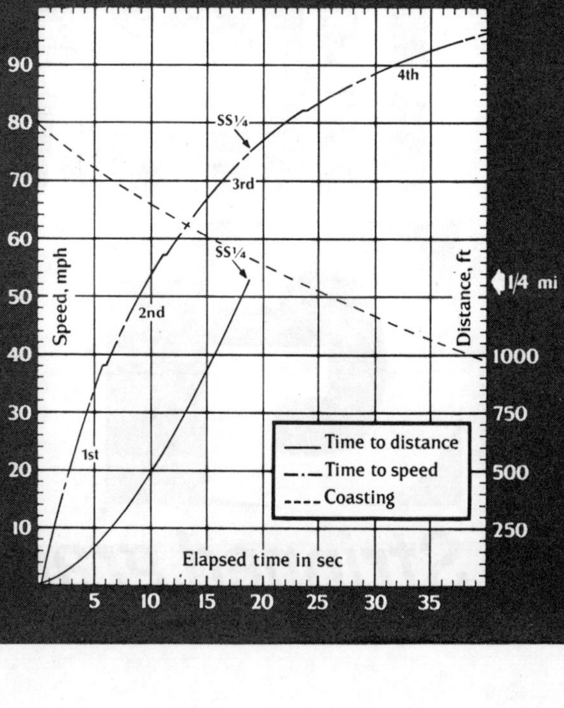

ABOVE: Rear shot emphasizes car's

TRIUMPH GT6

Stripped grand tourer

ROAD TEST

Six-cylinder lightweight Triumph is fast, smooth and eager, but watch the handlng . . .

GT 6 interior is smart and comfy — but equipment list is rather thin.

BE warned. The Triumph GT 6 suffers from the sudden onset of the dreaded swing-axle oversteer. You must keep the power on through a corner or the back end acts like it is trying to nestle under an armpit.

This, of course, is the basic fault of the swing axle. Camber increases and track narrows as weight is thrown on the front wheels. In a straight line this does not matter much — at least with the GT 6 — but when side forces are set up tyre contact area is reduced and resistance to them is therefore lowered.

If you know about all this, the new little Triumph grand tourer can be a delight.

The way to take a corner follows the old maxim of go in slow and come out fast. Keep the back wheels driving all the way in, and with a little practice you get to be able to turn it on well before the apex in most cases and really motor through. Just a question of technique — and just as fast as a lot of cars of this sort.

But what sort of a car precisely is the Triumph? It has high performance, a high price and not much equipment to go with it. A sort of stripped grand tourer.

Basically it is a Spitfire body turned into a two-seat coupe by using the tail and top of the Le Mans Spitfires and a slightly uprated version of the six-cylinder 2000 engine.

A smooth state of tune, complemented by high gearing that allows the 95 bhp unit to work well within its 1998 cc capacity. The car strides through the four forward gear ratios in fine style. It managed 46 mph in first gear, 68 mph in second, 96 mph in third, and a corrected maximum of 106.3 mph in top without ever sounding even so much as busy. The acceleration figures in our data sheet speak for themselves. And fuel consumption of 31.2 mpg should do something to ease the jolt of a $3,598 price tag.

The gearbox has a short central change lever — very quick, light and short in movement. The synchros are right up with the fastest change. But all the lower gears sing with the abandon of a choir getting stuck right into the Hallelujah Chorus. Later models will probably have differently cut gear teeth designed to quieten things down a bit.

The clutch bit well and smoothly. The accelerator spring was too weak for precise control.

So there's surely GT performance, plus a petrol bonus. How about comfort and equipment?

The list of optional extras supplies a good part of the answer — electric overdrive, heater-demister, wire wheels, radio, luggage straps and SAFETY BELTS. They might at least throw the belts in.

Creature comfort

The two doors are wide. The bucket seats are nicely shaped and adjust fore and aft over a wide range. No effort or strain to sit at the wheel for long distances.

Every control is easy to reach (without a safety belt) and clearly marked.

BELOW: Body is Spitfire-based. BELOW RIGHT: The 1998 cc. six-cylinder engine develops 95 bhp, pushes only 17 cwt.

TRIUMPH GT6

ABOVE: Spitfire-based lines are enhanced by Le Mans top and tail. Car remains strictly a two-seater. LEFT: GT 6 cornering at speed. Car will hit 108, cruises happily at 95.

Big round rev counter and speedo dominate the padded dash, with fuel and water temperature gauges set in the middle. The speedo has a trip.

Parcels shelf beneath the dash has a padded edge, and the centre console that carries the gear lever is also padded.

Alloy-spoked wheel with plastic padded rim is adjustable through two inches up and down by a screw knob. It is also adjustable for reach, but a spanner is needed. No trouble in suiting yourself, though.

Behind the seats is a flat, carpeted luggage space that won't take much without blocking off the back window — an outside mirror is also an extra. Spare wheel is under the floor.

Luggage space is got at either by tipping the seat backs forward or opening the big lid that carries the back window — a true GT touch, the inside light comes on when the lid is opened.

Bonnet is of the Herald-Spitfire pattern, folding forward to reveal engine and suspension. A mechanic's honeymoon.

Positive crankcase ventilation is fitted. A connection between inlet manifold and rocker cover contains a valve that allows a fairly constant part-vacuum to be kept in the crankcase, drawing off blow-by gases and soot and feeding them back into the cylinders to be burned. This does three things:

- REDUCES crankcase fumes and, therefore, air pollution — while the engine is in reasonable order. Fitting one to a clapped-out smoke bomb won't stop it from carrying around its stinking private fog.

- STOPS carbon and the like in blow-by fumes from contaminating engine oil and promoting wear.

- STOPS minor non-pressure oil leaks from points like timing gear covers.

The valve control snaps closed if the engine backfires, to stop flame reaching the sump.

Legislation is in hand to make positive control devices like these compulsory on all new cars in 1970. All Holdens, including the Torana, are now coming off the lines with them. They should help to lengthen the engine life, particularly with cars that do a low mileage over a long period when contaminated oil can form a highly corrosive and gritty mixture (which is, incidentally, one reason why we're not happy with these lengthened periods between oil changes. Sure, the oil itself doesn't break down, but what about the blow-by muck mixing with it?)

Brakes are disc at the front and drum at the back, with no power assistance. But pedal pressures are reasonable, and the result never in doubt. No fade or swerving.

Thank goodness the GT 6 is fitted with tyres to match the performance — Dunlop SP 41 radials in the case of the test car. The suspension is well-tuned to them, road noise level low and no harshness in the ride at low speeds. And are they good in the wet, under acceleration or brakes!

In a nutshell . . .

What we like about the car is the way it is so eager, smooth and (apart from that gearbox) quiet at all speeds. It's a wonderfully swift car for highway overtaking, which has a continuous cruising capability of 90-95 mph.

And the engine is a man just doing a boy's job. It has to lug around only 17 cwt. (unladen), and two litres is much more than enough for that.

At 100 mph it is turning over at slightly less than peak power speed of 5,000 rpm, and piston speed is below the generally accepted critical figure for wear of 2,500 ft. per minute.

The chances are for an engine like this giving top performance for a very long time without a lot of tuning or more serious maintenance work.

Even so, apart from the deficiencies in equipment, one or two design faults should be put right:

Pedal arrangement could be tidier — pedals closer to the floor and set for heel-and-toe.

The two-speed screen wipers set to cover a bigger screen area and made to go faster on both settings.

A bigger inside mirror set on top of the dash instead of hanging down from the top of the screen and taking a slice out of the driver's vision to the left.

And, most important, some form of compensating spring to tie down that swing axle. The handling will not be to everyone's taste, and driving rapidly needs a lot of concentration.

‘. . . the lower gears sing with the abandon of a choir getting stuck into the Hallelujah chorus . . .’

ROAD TEST DATA SHEET— TRIUMPH GT6

Manufacturer: Standard - Triumph, Coventry, England.

Test car supplied by Australian Motor Industries, Cook Street, Pt. Melbourne.

Price as tested: $3,598, including sales tax.

SPECIFICATIONS

ENGINE

Water cooled, six cylinders in line. Cast iron block, four main bearings.

Bore x stroke	74.7 x 76.0 mm.
Capacity	1998 cc.
Compression	9.5 to 1
Carburettor	Twin side-draught Strombergs
Fuel pump	mechanical
Fuel tank	9.75 gallons
Fuel recommended	super
Valve gear	p'rod ohv
Max. power (gross)	95 bhp at 5000 rpm
Max. torque	117 lb/ft. at 3000 rpm
Specific power output	47.5 bhp/litre
Electrical system	12v, 38 amp. hr. battery, 300 watt generator

TRANSMISSION

Four speed manual all synchro gearbox. Single dry plate clutch.

Gear	Ratio	Overall	Mph/1000 rpm	Max. mph
Rev.	3.10	10.15	—	—
1st	2.65	8.66	7.6	46 (6000)
2nd	1.78	5.82	11.3	68 (6000)
3rd	1.25	4.11	16.0	96 (6000)
4th	1.00	3.27	20.15	106.3 (5250)

Final drive ratio .... 3.27 to 1

CHASSIS

Wheelbase	6ft. 11in.
Track front	4ft. 1in.
Track rear	4ft. 0in.
Length	12ft. 1in.
Width	4ft. 9in.
Height	3ft. 11in.
Clearance	5in.
Kerb weight	17 cwt.
Test weight	19.1 cwt.
Weight distribution front/rear	56/44%
lb/bhp	19 lb.

SUSPENSION

Front: Independent by wishbones and coil springs, anti-roll bar and telescopic shock absorbers.

Rear: Independent by swing axles with transverse leaf spring and radius rods. Telescopic shock absorbers.

Brakes: Disc/drum; 260 sq. in. of swept area.

Steering .... rack and pinion

Turns lock to lock 4.25

Turning circle .... 28 ft. between kerbs

Wheels: Steel disc with 155 by 13 tubed radial (Dunlop) tyres.

PERFORMANCE

Top speed	108.7 mph
Average (both ways)	106.3 mph
Standing quarter mile	18.8 sec.

Acceleration

Zero to	seconds
30 mph	3.2
40 mph	5.8
50 mph	8.2
60 mph	11.7
70 mph	14.9
80 mph	20.8
90 mph	32.4
100 mph	46.7

	3rd	Top
20-40 mph	6.1	—
30-50	6.5	7.5
40-60	6.8	7.6
50-70	7.1	8.1
60-80	8.8	10.2
70-90	12.0	14.1
80-100	—	20.4

Braking: Ten 50 lb. from 60 mph.

Stop	percent G	pedal pressure
1	.91	50 lb.
2	.90	50
3	.92	50
4	.86	50
5	.88	50
6	.60	50
7	.68	50
8	.60	50
9	.60	50
10	.60	50

Consumption: 31.2 mpg over 311 miles, including all tests.

Speedo error:

Indicated mph	30	40	50	60	70	80
Actual mph	29	39	48	58	67	76

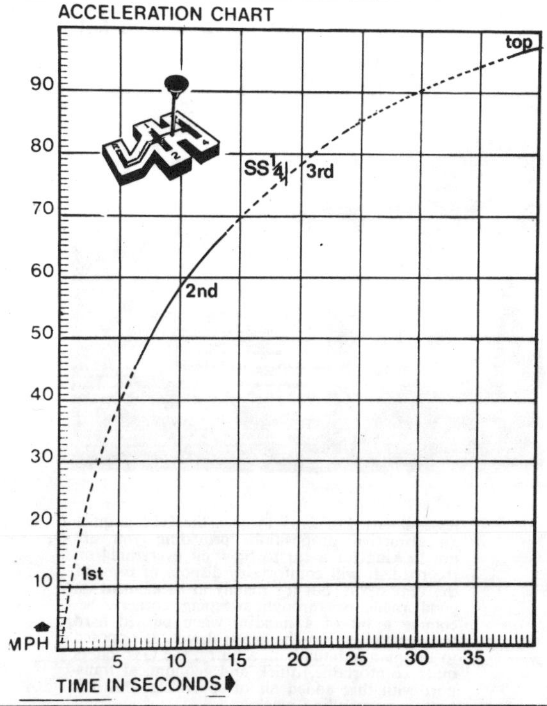

HOW TRIUMPH GT6 COMPARES

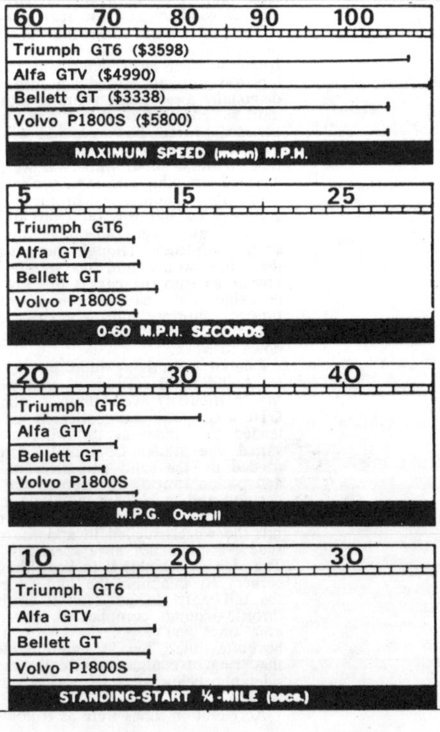

MOTOR RACING

CARS OF THE MONTH

Triumph GT6

Description: The GT6 is, as the name suggests, a grand touring, six-cylinder machine and is based closely on the earlier Triumph Spitfire. The backbone chassis itself is very similar, with the 2 litre engine as used in the Vitesse and Triumph 2000 models installed between branches at the front and the whole thing clothed in a coupé body very similar to that which appeared on the Le Mans Spitfires of 1965. General trim and luggage space is as befits a medium capacity 'grand tourer'.

Engine and transmission: The engine is a six-cylinder in-line unit of 1,998 cc; bore and stroke measurements being 74.7 mm x 76 mm. This produces 95 bhp at 5,000 rpm, and with twin Stomberg 1.50CD carburettors, is a very smooth and docile unit. Transmission is via an 8½ inch Borg & Beck diaphragm clutch to the four-speed constant mesh gearbox. This has synchromesh on all gears, and overdrive on third and top is offered as an optional extra for £58 7s 8d. A central gearchange is used, with a stubby lever falling easily to hand on the bulky transmission tunnel (which offsets the pedals to the right, giving a 'diagonal' driving position). Drive to the hypoid bevel rear axle is via swing axle shafts driven through needle bearing universal couplings. Final drive ratio is 3.27:1 or 3.98:1 with overdrive.

Suspension and brakes: As is normal Triumph practice, the GT6 has all independent suspension. At the front twin wishbones with co-axial coil spring/damper units are used. An anti-roll bar is fitted and rubber bushed wishbone pivots are used, together with patented screwed bottom bush and top ball joint swivels. At the rear the much-criticised swing axle independent system is retained, suspended by a transverse leaf spring and located by single radius arms on either side. Suspension loadings are fed into the double backbone closed channel section chassis frame, while the rear radius arms pick up on the underside of the body shell. The brakes are a combined disc and drum system, Girling all round, with 9.7 inch diameter discs at the front and 8 inch x 1¼ inch drums at the rear. A fly-off handbrake is mounted on the transmission tunnel, working mechanically on the rear drums.

Accommodation and fittings: The GT6 is strictly a two-seater with a large raised shelf providing extra luggage space only behind the seats. Access to this shelf, which is carpeted and lifts to reveal the spare wheel, is provided by a wide opening rear door, including the rear screen, which opens upwards on its spring-loaded hinges. The seats themselves are well shaped and quite comfortable; providing a lot of lateral support with raised hip pieces, if being rather on the narrow side for broad-in-the-beam drivers. A very bright interior lamp is fitted to the roof, and a handsome walnut facia is used, carrying quite an impressive, not to say confusing, array of switches and gauges. Behind the slightly dished leather-trimmed steering wheel are a speedometer and rev counter, and high to the left are the water temperature and fuel gauges, but no oil pressure gauge. Controls include wiper, lights, heater motor and ignition switches and washer, heater temperature and choke controls, while stalks on either side of the collapsible steering column are the overdrive switch, and an omni-purpose lighting control, giving main beam, dipped beam, side-lights and flash. A centre horn push is used. All foot pedals are of the pendant type, with long bowed accelerator pedal and a handy left-foot rest beside the clutch pedal.

The bonnet is of the usual fully-opening Standard-Triumph type, with retaining catches on either side behind the front wheel arches. This is an excellent feature, making the engine and its ancillaries most accessible.

Findings: Our test car was one of the first GT6s to have been built and as such had completed something over 16,000 miles by the time we got our hands on it. This makes our test something of a 'used car review', but to try a car like this was quite interesting. Prospective buyers of a used GT6 should examine the door sills and body seams for signs of rust, for 'ours' was going there quite badly.

However, first impressions of the car were quite good. There is not enough adjustment in the driver's seat to make it a really comfortable car for anyone slightly above average height, but the GT6 gives a pleasantly solid feeling: it's one of those cars that you 'put on' rather than 'sit in'. Controls fall easily to hands and feet, but a bad point is that no overdrive warning light is included, and it can be a bit confusing not knowing whether it's engaged or not. Forward vision is good over the long bonnet with its 'power bulge' in the middle to clear the six's cylinder head, and rear mirror vision through the large screen was also good. But reversing the GT can be difficult for there are large blind spots between the part-opening rear quarter windows and the rear screen itself.

The gearbox was surprisingly stiff for a 16,000-miler, and often it was difficult to engage first from rest. The lever movement from third into top was also an odd down into neutral, then diagonally away from yourself into fourth, and until we got used to this it was quite slow. Experience proved, however, that when you really needed the gears in a hurry, and didn't have time to think about what you were doing, they engaged smoothly and rapidly almost every time —and this although synchromesh on third was getting tired and was far from unbeatable.

The brakes were adequate, but didn't give great confidence, Though there was no doubt that they would pull up square, they didn't appear to stop the car as quickly as might be desirable. This may have been purely a question of adjustment on the test car, though, since the pedal had to be pressed quite a long way down to work.

One of the driving hints in the handbook is to 'Avoid rapid cornering'—and we can add to this 'particularly over bumpy surfaces . . .'! The GT6 was quite badly affected by bumps and tended to wander as its swing axle back end varied the track. Continual corrections were needed on the solid and pleasant feeling rack and pinion steering to keep to the straight and narrow, and in corners initial understeer would scrub off a little speed before the tail began to lift, the wheels tucked in and the tail began to slide . . . and not always exactly controllably. The idea (we think!) is to keep the power on in corners to minimise the understeer, and once the tail starts to go keep it in hand on the throttle—sounds complicated, doesn't it? However, once you've got the knack, the GT6 can be quite quick, but its handling does not give that 'ring of confidence'. Really its limit is considerably below that of several other, similar vehicles.

Acceleration times were as follows:

0-30 mph— 3.7s
0-40 mph— 6.0s
0-50 mph— 8.7s
0-60 mph—13.6s
0-70 mph—16.1s
0-80 mph—22.0s

The usual 300-plus test mileage returned fuel consumption figures of 24.5 mpg, which were very good for a 2 litre six being pushed through the gears quite regularly.

Assessment: For £985 1s new, the GT6 is quite an attractive proposition, providing you are not looking for a car to blow off everything on the road. It will comfortably dispose of most of the opposition, but it's mainly in its element on good roads or through sweeping corners, becoming a bit of a handful when pushed hard through twists. For the gentle driver, as opposed to the man habitually in a hurry, the GT6 offers quite comfortable, quick and economical transport with that added air of sportiness. But it is not an outstanding machine.

Triumph GT6 1,998 c.c.

AT A GLANCE: Spitfire derivative with Triumph 2000 engine. Very quiet and smooth; excellent performance. Progressive brakes. Rear suspension needs improving. Varied roadholding, very sensitive to the throttle. Poor ventilation, restricted visibility and cramped interior. Luxury finish at reasonable price.

MANUFACTURERS:
Standard-Triumph International Ltd., Coventry, Warwickshire.

PRICES:

Basic (with heater) ..	£811 0s 0d
Purchase Tax ..	£187 11s 5d
Seat belts	£7 19s 9d
Total (in G.B.) ..	£1,006 11s 2d

EXTRAS (inc. P.T.)

Overdrive	£58 7s 8d

PERFORMANCE SUMMARY

Mean maximum speed	106 m.p.h.
Standing start ¼-mile	18·5 sec
0-60 m.p.h.	12·0 sec
30-70 m.p.h. (through gears)	12·0 sec
Fuel consumption ..	24 m.p.g.
Miles per tankful ..	230

IT has taken almost a year for Standard-Triumph to get around to letting us test their exciting new GT announced for the London motor show in 1966. Initially the car was for export only, but from January this year supplies have been coming through on the home market. In many ways this Triumph is unique because it comprises a formula never offered before at an all-in price so close to £1,000.

Essentially the GT6 is a Spitfire sports car, transformed in character by fitting the engine from the luxury Triumph 2000 saloon. The body is modified from the open two-seater into an elegant fastback coupé and the interior is completely restyled to suit the outward appearance.

Mechanically the chassis is identical with that of the Spitfire, with the steel body bolted up on a separate double backbone. Independent suspension is used front and rear and there are front disc brakes working without servo assistance. The engine is slightly different from that of the 2000 saloon in that the compression ratio is 9·5 instead of 9·0 to 1, which puts the net power output up from 90 to 95 b.h.p. The gearbox is a " special " for the GT6 and Vitesse 2-litre, being essentially a beefed-up Herald unit with synchromesh added on bottom.

Compared with a Spitfire, its performance is improved out of all recognition because the new engine adds 42 per cent more power. Our latest test of a Spitfire is the Mk. 2 version and since then there has been a slight increase in capacity, but referring back to that test of 26 August 1966 shows the GT6 to be 14 m.p.h. faster on top speed and 12·4sec quicker in reaching 80 m.p.h. from rest.

These two statistics hardly begin to tell the story though, because it is the extra torque of the 2-litre six which really counts; from 20 to 80 m.p.h. in top gear, for example, takes half as long in the GT6 as in the Spitfire. Apart from the times recorded, the 6-cylinder engine is much smoother and quieter than the four and because it does not like to rev as much, the gearing has been raised to give the car a more easy-going and relaxed gait.

As usual, overdrive on third and top is an extra for £58 and the test car had it fitted. A much lower final drive ratio (3·89 instead of 3·27 to 1) is used in conjunction with overdrive, so that the highest gear m.p.h. per 1,000 r.p.m. is not very different for the two versions. Acceleration is probably quicker with the overdrive car because the non-overdrive one has very high indirects; bottom, for ex-

Make: Triumph
Type: GT6
1,998 c.c.

TEST CONDITIONS
Weather: Sunny. Wind 10-15 m.p.h.
Temperature: 20 deg. C. (68 deg. F.)
Barometer: 29·20in. Hg.
Humidity: 50 per cent
Surfaces: Dry concrete and asphalt

WEIGHT
Kerb weight: 17·5cwt (1,964lb-891kg)
(with oil, water and half-full fuel tank)
Distribution, per cent: F, 57·4; R, 42·6
Laden as tested: 20·2cwt (2,25/lb-1,045kg)

Test distance 1,142 miles. Figures taken at 17,000 miles by our own staff at the Motor Industry Research Association proving ground at Nuneaton.

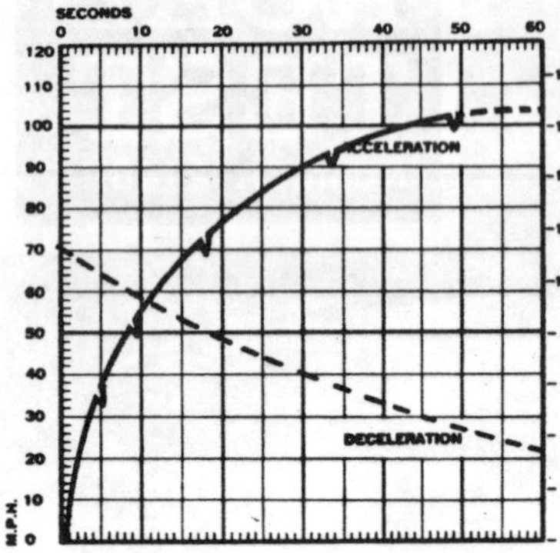

MAXIMUM SPEEDS

Gear	m.p.h.	k.p.h.	r.p.m.
O.D. Top (mean)	106	171	4,900
(best)	108	174	5,000
Top	100	161	5,780
O.D. 3rd	98	158	5,670
3rd	83	134	6,000
2nd	58	93	6,000
1st	39	63	6,000

Standing ¼-Mile 18·5 sec 75 m.p.h.
Standing Kilometre 34·3 sec 92 m.p.h.

FUEL CONSUMPTION

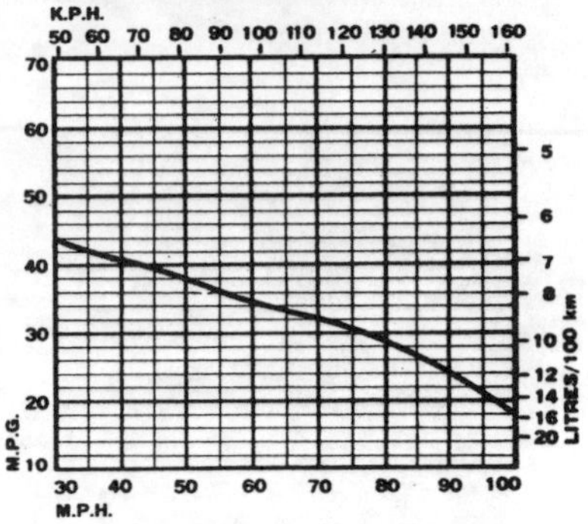

(At constant speeds in O.D. Top—m.p.g.)
Speed	m.p.g.
30 m.p.h.	43·1
40	40·8
50	38·1
60	35·0
70	31·5
80	28·4
90	24·8
100	17·3

Typical m.p.g. 24 (11·8 litres/100km)
Calculated (DIN) m.p.g. 28·6 (9·9 litres/100km)
Overall m.p.g. 20·2 (14·0 litres/100km)
Grade of fuel, Super premium 5-star (min 100 RM)

OIL CONSUMPTION
Miles per pint (SAE 10W/30) .. 750

TIME IN SECONDS	3·6	5·9	8·5	12·0	15·6	21·2	31·3
TRUE SPEED M.P.H.	30	40	50	60	70	80	90	100
INDICATED SPEED	31	41	51	61	70	80	90	100

Mileage recorder 4 per cent over-reading.

Speed range, gear ratios and time in seconds

m.p.h.	O.D. Top (3·11)	Top (3·89)	O.D. 3rd (3·89)	3rd (4·86)	2nd (6·92)	1st (10·31)
10—30	—	—	—	6·1	4·2	3·2
20—40	9·4	6·7	7·4	5·9	3·9	3·6
30—50	9·2	6·9	7·1	5·8	4·5	—
40—60	10·3	7·9	8·0	6·2	—	—
50—70	11·5	8·5	8·7	7·5	—	—
60—80	14·6	10·0	10·4	11·6	—	—
70—90	19·6	16·1	17·7	—	—	—

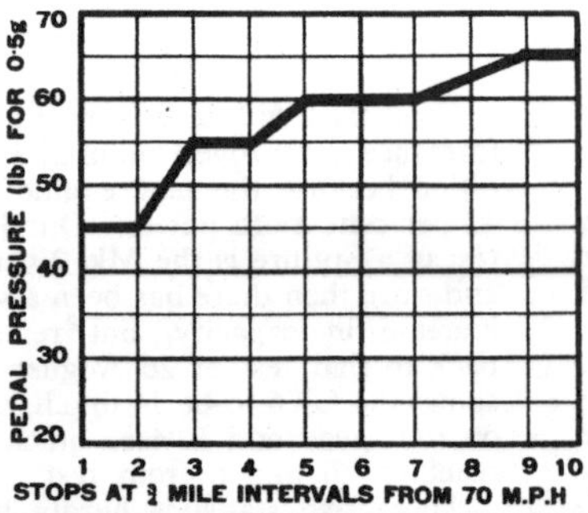

BRAKES (from 30 m.p.h. in neutral)

Load	g	Distance
25 lb	0·25	120ft
50 "	0·40	75 "
75 "	0·70	43 "
100 "	1·05	28·6 "
125 "	1·10	27·3 "
Handbrake	0·30	100 "

Max. Gradient, 1 in 3
Clutch Pedal: 20lb and 4in.

TURNING CIRCLES
Between kerbs L, 25ft 2in.; R, 25ft 4in.
Between walls L, 26ft 9in.; R, 26ft 11in.
Steering wheel turns, lock to lock 4·3

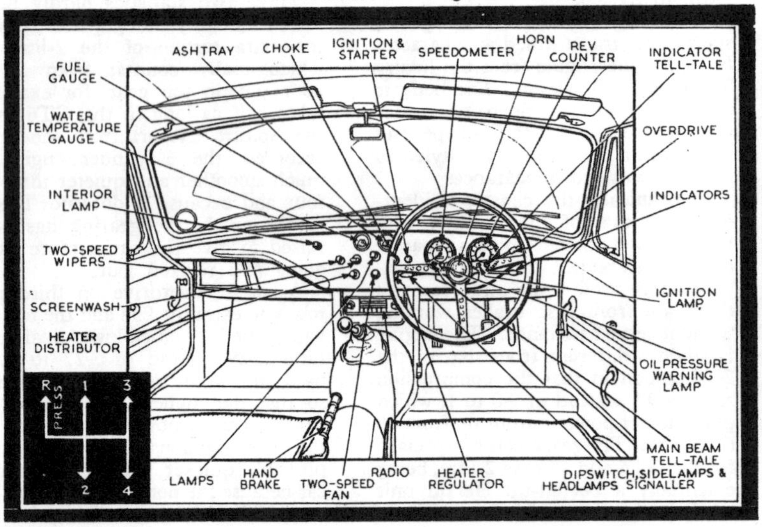

HOW THE CAR COMPARES:

MAXIMUM SPEED (mean) M.P.H.
Triumph GT6
Triumph TR4A
MGB GT
Alfa Romeo Sprint GTV
Lotus Elan Coupe

0-60 M.P.H. (sec)
Triumph GT6
Triumph TR4A
MGB GT
Alfa Romeo Sprint GTV
Lotus Elan Coupe

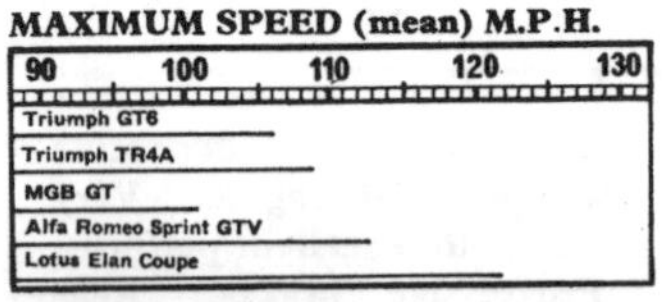

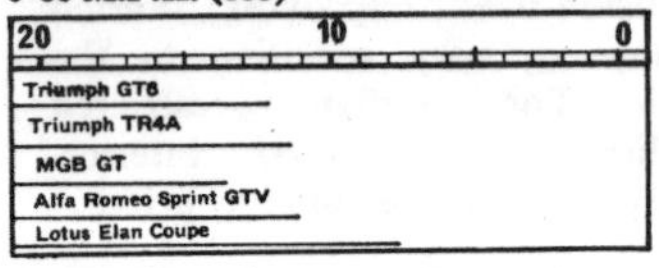

STANDING START ¼-mile (sec.)
Triumph GT6
Triumph TR4A
MGB GT
Alfa Romeo Sprint GTV
Lotus Elan Coupe

M.P.G. OVERALL
Triumph GT6
Triumph TR4A
MGB GT
Alfa Romeo Sprint GTV
Lotus Elan Coupe

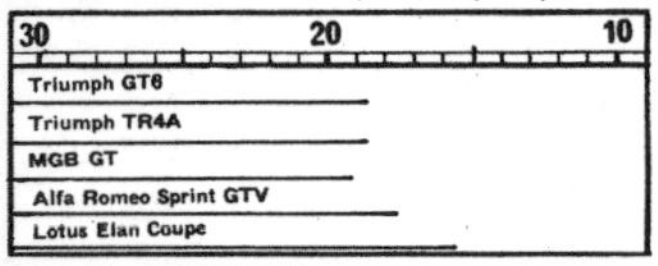

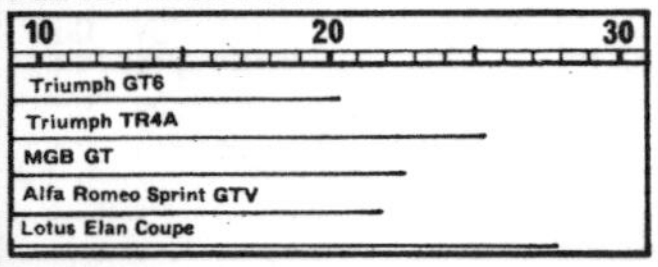

PRICES
Triumph GT6	£985
Triumph TR4A	£985
MGB GT	£1,065
Alfa Romeo Giulia Sprint GTV	£1,950
Lotus Elan Coupe	£1,598

Seats are softly padded and covered in stretch Ambla with good wrap-round support. The two main instruments are directly in front of the driver

ample, would run to 48 m.p.h. on the non-overdrive car.

Although the overdrive third ratio is almost identical with direct top, we found there was a slight advantage during the acceleration runs in using every gear available and working up through the composite system as though it were a 6-speed box. Take-offs were dramatic, with the back of the car squatting down and the wheels spinning easily on dry concrete. On the road, however, the high gearing kills this tendency, unless the surfaces are wet, and the GT6 simply rears up and goes away from traffic lights and hold-ups with a long, hard and very quiet surge of power.

The car is decidedly sensitive to all its power and the bonnet can be seen heaving up and down in proportion to the throttle opening. At night the headlamps swing up and down even more and other traffic seemed to be troubled by our beams whenever we were accelerating. Small drivers sitting well back also found the forward view quite seriously reduced as the front came up.

Roadholding

This characteristic naturally affects the rear swing axles, which change track and wheel camber as the suspension goes up and down. In corners with power on, the GT6 tries to understeer up to the point where speed and torque are sufficient to break the tail away. Lifting off in a corner causes the back of the car to rise as the weight transfers forwards and the rear wheels then take on positive camber with a narrower track between them. This reduces their cornering power considerably and the tail swings out immediately.

There are, therefore, three distinct patterns of cornering behaviour and the car feels unsteady and difficult to control during the transitions between each of them. If the driver makes up his mind how he is going to tackle a particular bend well in advance and sticks to his plans, all is well. If some emergency arises during cornering which causes him to lift off or change his line suddenly, then the car could get out of hand, particularly on a slippery surface.

We had no real trouble of this type during our 1,142-mile test, but we found the GT6 very difficult to drive fast tidily, particularly through roundabouts and along twisty lanes. The steering is accurate but rather too low geared, and although the rubber mountings for the rack effectively take out all the road shock, they do introduce some free play and lack of precision. On bumpy roads the rear wheels also have a tendency to steer the car slightly and the driver feels he can never relax his concentration on keeping the car straight.

Movements of the rear suspension are restricted, so there is a lot of violent bottoming on bad roads and on the MIRA *pavé* track the little coupé had a very rough and noisy time. The outside rear wheel bottoms even during hard cornering on a smooth road, making an alarming "bonk, bonk, bonk" noise under the floor. We feel it is a pity that the limitations from the continued use of swing axles should detract so much from what is basically such a good car, and we urge Standard-Triumph to make improvements without delay.

The brakes work very well, with a progressive response to increasing pedal load. At only 100lb we recorded over 1g without any wheel lock or signs of slewing, and 125lb increased the retardation even farther, so that the column of fluid in our Mintex decelerometer disappeared completely off the scale. The fade test caused a 50 per cent rise in pedal load after 10 stops from 70 m.p.h. with a lot of hot smells. The handbrake must now be the last remaining example of fly-off type; it proved effective when used on its own from 30 m.p.h. and held securely on a 1-in-3 hill facing either way. Restarting was no problem either.

Fuel consumption depends very much upon the cruising speeds chosen, the GT6 using petrol twice as fast at 100 m.p.h. as it does at 60 m.p.h. Overall our figure was slightly better than 20 m.p.g., but on quite a fast trip from London to Somerset and back in an afternoon we recorded 28 m.p.g. Driven sensibly, with as

The Triumph 2000 engine looks long under the Spitfire bonnet and it is given a chromium plated rocker box. Everything is very accessible

Fastback lines suit the little coupé very well. There are louvres on top of the bonnet each side of centre to help with the cooling

Triumph GT6 . . .

much use of the overdrive as possible, the car should prove very economical. Super premium 5-star fuel is required.

The overdrive is worked by a stalk on the steering column, which looks just like the other two for lamps and indicators. Engagement is jerky unless the engine is pulling hard at quite high revs, so it is better to stab the clutch pedal at just the right moment. Overdrive top will pull without snatch from as low as 20 m.p.h., but on the test car the unit then groaned quite loudly right through to about 50 m.p.h. It was always noisy when in use and seemed to accentuate a prop shaft vibration at about 4,500 r.p.m. in top.

Wind noise is low with all the windows shut tight, but then there is no ventilation at all. The rear quarter panes click open on over-centre catches, but their extraction effect is very slight and we found no way to keep ourselves from getting sticky in warm weather except by creating a great blast of air from a front quarterlight. The heater has very crude controls by today's high standards, with no face level ducts and just a water valve temperature control pulling out of the facia like a choke knob and another matching one for the distribution to screen or footwells. The booster fan has two speeds.

The interior of the GT6 is confined and anyone over about 5ft 8in. tall or 40in. round the hips cannot get comfortable. The seats are well shaped to grip the bottom and back and they hold the occupants very well against cornering forces, with good lumbar support and a comfortable fixed angle of rake. The interior door handle is squashed in between the seat cushion and the door where it is hard to find and clumsy to use. The window winder is out of reach at the farthest part of each turn if the seat belts are done up tight.

Compared with the Spitfire's, the facia layout has been completely revised with a new polished veneer panel curving down in the centre of the car to join up with the console. Large matching dials for the speedometer and rev counter come directly under the driver's eyes with the fuel gauge and engine thermometer in the centre of the car. Switches are an odd collection of toggles and knobs, but each has its function displayed by a symbol and their positions are in fact sensible and logical. Somehow the smart leather-covered Formula steering wheel seems large, although it is only 15in. diameter. There is a good sturdy, two-handed grab handle in front of the passenger.

Pedals are well placed and there is a rest for the left foot alongside the clutch. The throttle had a jerky take up sometimes on the test car and it was prone to rattle through its linkage against the driver's foot. Heel-and-toeing of the brake and throttle at once is easy.

The gearbox is sticky rather than stiff to use, with notchy movements through the slightly skew gate. There is a push-down guard for reverse which is alongside first, but because of the angle at which the lever is cranked it is possible to over-ride this by mistake if one is too hasty or clumsy. The synchromesh is only just able to take fast changes without crunching. There was some faint whine from the indirects on the test car.

As mentioned earlier, forward vision depends on the angle of the bonnet and it is very hard to see the raised edges of the front wings, especially as the wipers park across that crucial lower part of the screen. The framed interior mirror is too small to make the most of the large rear window and an outside mirror would be an advantage. The blind rear quarters give no real trouble in traffic, but they do make it difficult to line up when reversing through a narrow gateway. Tall drivers find the car lacking in headroom, with a beetle-browed effect at the top of the screen.

With the seat cushion no more

Spring-assisted hinges for the tailgate are well concealed. Spare wheel and tools are under the floor of the rear compartment

than about 10in. from the road surface only the agile should try to leap quickly into the GT6. The edge of the bucket seat is hard and usually the first thing encountered.

The big back window lifts up on concealed, spring-assisted hinges to make it very easy for loading luggage on to the large rear platform. There are two slots in the floor and carpet, presumably for straps to secure the cases. Unfortunately all one's belongings must be exposed to view (a hard temptation for thieves to resist in some countries) except for the few tiny things one can insinuate into two boxy compartments behind the seats. The parcel shelf each side under the facia is open to view and really only deep enough to take maps and a touring guide.

Overall this report is critical, and has been made only after much discussion. Our test staff were quite divided in their opinions, some feeling very strongly that the rear suspension and lack of ventilation spoilt a potentially great machine, others managing to live with its limitations and enjoy the other very worthwhile virtues in the car. The opinions of the bigger testers were further coloured by their discomfort, so as a team it is difficult to be objective. Potentially the GT6 is a fine formula; with further development (and if necessary a price increase) it could become outstanding. ■

SPECIFICATION: TRIUMPH GT6 (FRONT ENGINE, REAR-WHEEL DRIVE)

ENGINE

Cylinders ..	6, in line
Cooling system ..	Water; pump, fan and thermostat
Bore ..	74·7mm (2·94in.)
Stroke ..	76·0mm (2·99in.)
Displacement ..	1,998 c.c. (122 cu. in)
Valve gear ..	Overhead, push-rods and rockers
Compression ratio	9·5-to-1
Carburettors ..	2 Stromberg 150 CD
Fuel pump ..	AC mechanical
Oil filter ..	Full flow, renewable element
Max. power ..	95 b.h.p. (net) at 5,000 r.p.m.
Max. torque ..	117 lb ft (net) at 3,000 r.p.m.

TRANSMISSION

Clutch ..	Borg and Beck 8·5in., diaphragm spring
Gearbox ..	4 speed, all-synchromesh
Gear ratios ..	O/D Top 0·80, Top, 1·00; O/D Third, 1·01, Third, 1·25; Second 1·78; First, 2·65; Reverse 3·10
Final drive ..	Hypoid bevel 3·89 to 1

CHASSIS and BODY

Construction ..	Separa testeel backbone chassis, welded steel body

SUSPENSION

Front ..	Independent, coil springs, wishbones, anti-roll bar, telescopic dampers
Rear ..	Independent, swing axles, transverse leaf spring, radius rods, telescopic dampers

STEERING

Type ..	Alford and Alder, rack and pinion
Wheel dia. ..	15in.

BRAKES

Make and type ..	Girling, disc front, drum rear
Servo ..	None
Dimensions ..	F, 9·7in. dia.; R, 8in. dia. 1·25 in. wide shoes
Swept area ..	F, 197 sq. in.; R, 63 sq. in. Total 260 sq. in. (258 sq. in. per ton laden)

WHEELS

Type ..	Pressed steel disc, 4-stud fixing, 4·5in. wide rim. Optional wire spoke wheels
Tyres—make ..	Goodyear
—type ..	G800 radial ply
—size ..	155-13in.

EQUIPMENT

Battery ..	12-volt 48-amp. hr.
Generator ..	Lucas C40L d.c.
Headlamps ..	Lucas sealed beam, 90-120 watt (total)
Reversing lamps	2, standard
Electric fuses ..	3
Screen wipers ..	Two-speed, self-parking
Screen washer ..	Standard, manual plunger
Interior heater ..	Extra, water valve control
Safety belts ..	Extra, anchorages built in
Interior trim ..	Ambla seats, pvc headlining
Floor covering ..	Carpet
Starting handle	No provision
Jack ..	Screw pillar
Jacking points ..	Under chassis
Other bodies ..	None

MAINTENANCE

Fuel tank ..	9·75 Imp. gallons (no reserve) (44·3 litres)
Cooling system ..	11 pints (including heater) (6·2 litres)
Engine sump ..	8 pints (4·5 litres). SAE 10W/30 Change oil every 6,000 miles; change filter element every 12,000 miles
Gearbox ..	1·5 pints SAE EP75 No change; check level every 12,000 miles
Final drive ..	1 pint SAE EP90. No change; check level every 12,000 miles
Grease ..	2 points every 6,000 miles 2 points every 12,000 miles
Tyre pressures	F, 20; R, 24 p.s.i.

PERFORMANCE DATA

Top gear m.p.h. per 1,000 r.p.m.	17·3
Overdrive top m.p.h. per 1,000 r.p.m.	21·1
Mean piston speed at max. power	2,495ft/min
B.h.p. per ton laden	94

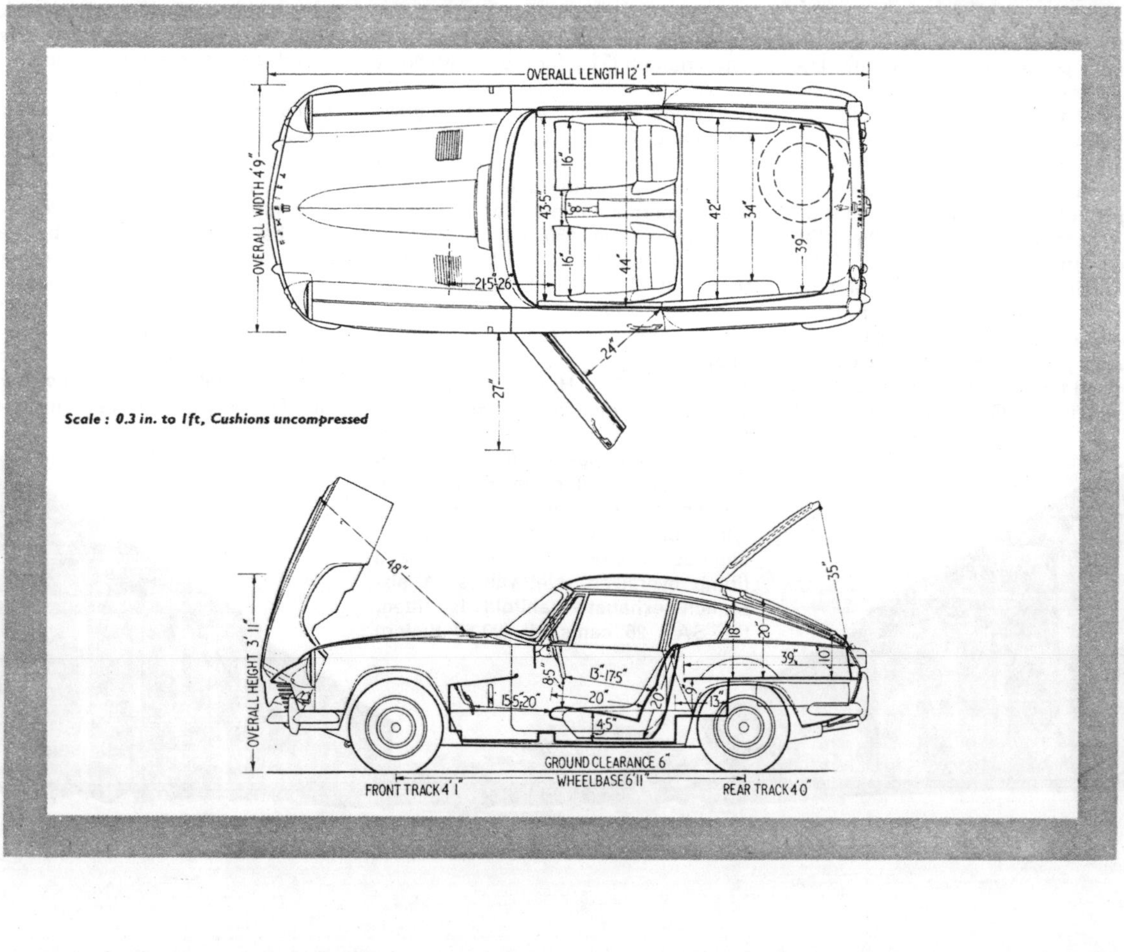

Scale : 0.3 in. to 1ft, Cushions uncompressed

SAH GT6 STAGE 3 *Fuel Injection*

WHEN the finished product is such a desirable property it's difficult to understand why only one firm, really, specialises in the large-scale radical tuning of Standard-Triumph products. The sporting Triumphs are really very satisfactory as they stand, but none of the big five has yet turned out a car which wasn't the better for a little skilled attention here and there, if only to make your Bloggsmobile a bit different from all the others in the road. When the skilled attention is given by SAH Accessories, of Linslade, Leighton Buzzard, Beds., the finished product is likely to be a very individual fast car, as we've just discovered in the course of a road test of Syd Hurrell's Stage 3 tuned Triumph GT6. Quite apart from anything else (and there's a lot else) this car has Tecalemit-Jackson/SAH fuel injection, a scheme which gets rid of carburettors and which is all the go these days.

The Triumph GT6, for those who don't actually have the information at their finger-tips, is a sort of highly-developed Spitfire (i.r.s., two seats and so on) with a six-cylinder Triumph 2000 two-litre go-department; in standard form this has a 9.5 to 1 compression ratio, two 150 CD Stromberg carbs and squirts out 95 b.h.p. at five thousand. Top whack is about 120, and it gets from rest to 60 in just under 10 seconds, and from rest to 80 in about 17½.

By the time Syd Hurrell has finished with it, the engine is still two litres, but this is about all you can say. The cylinder head has been reworked to SAH Stage 2, which involves raising the compression ratio to 10.5 to 1, modifying, balancing and polishing the combustion chambers and ports, and fitting oversized inlet valves. A six-branch exhaust manifold is fitted, the SAH 26 camshaft (20-60 timing) goes in, with special valve springs, and the carburettors are, of course, replaced by the TJ fuel injection equipment. Then there's an oil cooler and a Lucas sports coil.

On "our" car, even this was not all. We got a car with a twin-silencer assembly, Armstrong adjustable dampers front and rear, Minilite 5½J mag-alloy wheels with 165 x 13 Cinturatos and spacers at the front, and a very smart glassfibre bonnet section called the "SAH Le Mans", which fairs-in the headlights, does away with the standard GT6 "power bulge" and adds a set of louvres to the scuttle sides to help keep down the underbonnet temperature. Inside the car, the driver's seat mountings are altered to give more rake to the seat backrest and, on the car we tried, an oil temperature gauge, plus an additional oil temperature gauge, were fitted on the dash, plus a control for the Kenlowe electric fan.

The replacement bonnet probably doesn't have much aerodynamic effect but it certainly saves a fair-sized slab of weight (around 40 lb.) by doing away with an enormous steel pressing; the engine mods are reckoned to give an increase of 50-odd b.h.p. which means that there's a fair bit of extra power to put on the road, although we had no wheelspin even when taking performance figures, for which purpose we found it best to drop the clutch in at about 5,000 r.p.m. A hotter camshaft and big inlet valves usually suggest top-end power—and don't run away with the idea that the SAH car hasn't got any of this, 'cos it has. But it might also suggest a lack of flexibility, which is where the petrol injection affair starts to do its stuff. We've explained many times before in these pages that the big advantage of fuel injection is not that it gives an instant increase in urge, but that it makes it practical to take advantage of other mods which do. This is what happens in the case

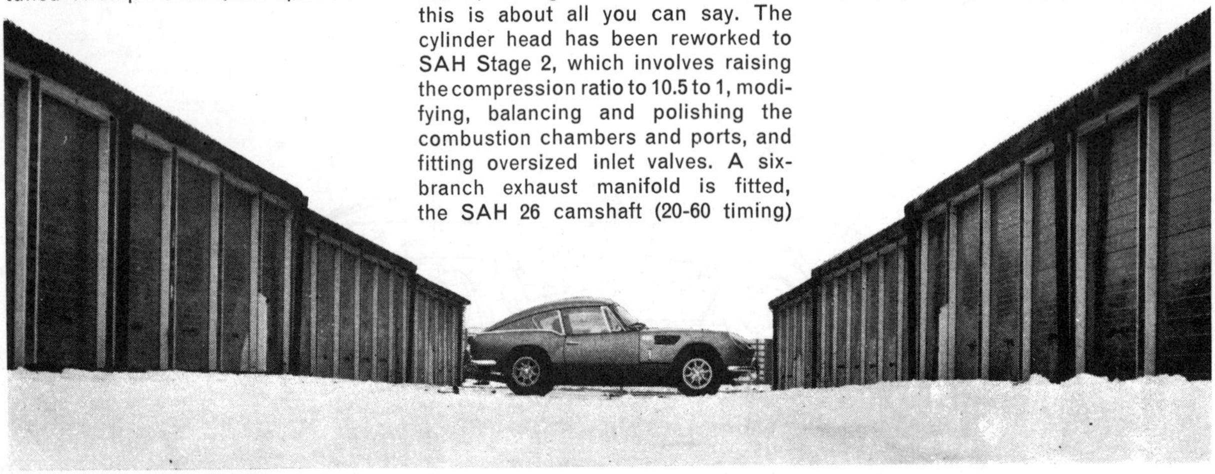

of the GT6; the combination of valve timing and valve size would ordinarily mean a car with a distinct distaste for proceeding at low r.p.m. with smooth behaviour, but the flexibility imparted by the injection arrangement gets rid of all this, and if you don't believe us, try banging the loud pedal wide open from about 1,200 revs. in overdrive top! The results are rather startling, and in fact the engine pulls smoothly from lower speeds than this in the same highest-available ratio. This isn't practical because the tickover speed on "our" car was 1,200 r.p.m. and obviously it would not run smoothly in or out of gear below this speed.

In the bleak mid-winter—the SAH "Le Mans" front end gives the GT6 a fresh approach, as you might say.

In the engine room—picture shows the fuel injection set-up, which is why there ain't no carbs!

The same flexibility showed up in the middle-range, too, where acceleration in top or overdrive top from, say, 65-70 miles an hour (3,000 r.p.m. in overdrive top) was similar to the effect of trying the same thing in the standard GT6 in, say, third gear. Now yer see it, now yer don't.

For most people—certainly most enthusiastic drivers—this business of tooling about at zero revs in high gears is largely academic, and what interests most of us most of all is how the thing steps off. As we indicated at the beginning of this article, the standard GT6 ain't no slouch, so that any improvement is going to add up to a very sharp old motor indeed. With the SAH car, you reach 90 in the same time as the standard product gets to 80, and 80 comes up in the standard car's 70 time. And in the lower registers, you have a car which rockets to 60 from a standstill in under nine seconds, with no wheelspin or ungentlemanly fireworks. You take the revs to around "five", drop in the clutch and apply right foot, and suddenly you're a quarter of a mile down the road. Fabulous. Actual top whack is a bit unimportant in these Castellated days, but you can say that the SAH conversion puts the best part of ten miles an hour on to the standard GT6 maximum of 110 or so.

With this much performance the car obviously has to handle alright. We were able to examine this in some detail because the Met. Office had the kindness to lay on rain, freeze and snow, in quick succession, within a couple of hours of the car coming into our possession. The standard car has pretty good handling characteristics within the limits of swing-axle i.r.s., of course, and we were interested to see that apart from the items mentioned above—wide rims, fat tyres and adjustable dampers—the SAH boys had left it alone. Obviously, if you use too much urge on ice then you're going to get wheelspin and a wagging tail, but the car is always controllable and on better adhesion the extra urge is never an embarrassment. In fact, you could sum this car up as a very desirable high-speed touring carriage indeed.

PERFORMANCE FIGURES

(Figures in brackets relate to standard Triumph GT6)

Maximum speed (approx.) 120 m.p.h. (110)

Acceleration:	0-30	3.2 (4.0)
	0-40	4.9 (5.9)
	0-50	6.6 (7.7)
	0-60	8.4 (9.6)
	0-70	10.8 (13.5)
	0-80	13.4 (17.4)
	0-90	17.4 (21.8)

Fuel consumption: 22 m.p.g. overall (26 m.p.g.).

Cost of engine conversion, £238. Mag-alloy wheels and spacers, £69. Le Man's bonnet (unsprayed) £53. Extra gauges, £70. Kenlowe fan £17.11.6. Armstrong adjustable dampers £21.18.0. Dual silencer assembly, £9.15.0.

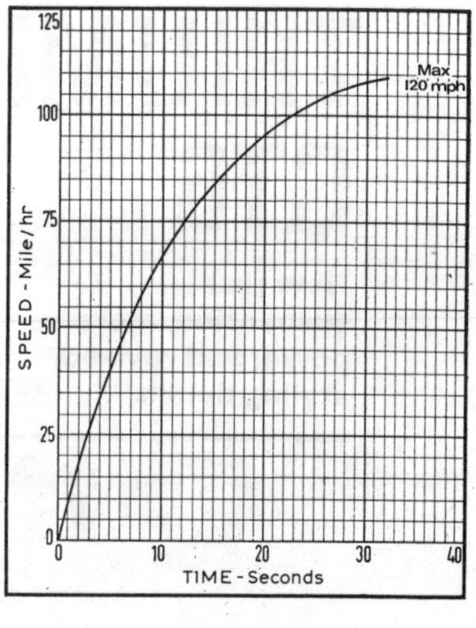

NEW FOR '69

MK 2 TRIUMPH GT6

New rear suspension, more power and revised styling details

INTRODUCED 2 years ago, and based on the Triumph Spitfire, the GT6 receives important changes for 1969. Old swing axle i.r.s. now replaced by wishbone layout, but transverse leaf spring remains, doubling as "top wishbone"; half shaft now has rubber doughnut in centre to accommodate plunge. More power through fitting TR5 head and new camshaft. Revised interior, facia, and external styling details. Heater now standard along with heated rear window.

THE first Triumph GT6, announced just two years ago, came in for heavy Press criticism because of odd handling caused by its swing axle i.r.s., and for unsatisfactory ventilation compared with other models. This week a revised Mark 2 version is announced which takes care of both these complaints and is in full scale production already. In addition there is a 9-bhp power increase, a completely re-styled facia and switches, and minor styling re-touches to the exterior.

Most important is undoubtedly the new rear suspension. Originally the Spitfire-Vitesse-Herald swing axle i.r.s. had been used, whose main virtue was simplicity, but whose main vice was a very high roll centre coupled with pronounced wheel camber changes between bump and rebound—more noticeable when lifting-off or tramping on the throttle which cornering under stress; soft damping to suit the GT6 to American tastes did not help either.

A long hard look at the rear of the car showed that there was space for a suspension revision, and the first new type tried was a coil-spring-cum-strut i.r.s. not unlike the Lotus Elan layout. This worked well, being raced with some success by Bill Bradley in a factory-backed car, but was rejected on the grounds of cost and considerable chassis and bodywork changes which would have been necessitated by the high strut. The final solution was really rather ingenious. The geometry has been completely changed from simple swing axle to double wishbone, but none of the original chassis frame has been changed, and the original transverse leaf spring and differential casing have been retained. Bearing many resemblances to single-seater racing practice, there is now a "reversed" cast bottom wishbone (i.e. with single pivot *inboard*) linking the new hub carrier to the existing chassis frame, and the transverse leaf spring doubles as a top "wishbone" rather like early Coopers. A re-aligned radius arm runs forward from the bottom of the hub carrier to pick up on the body much nearer the

PRICES

Basic	£879	0s	0d
Purchase Tax (in G.B.)	£246	5s	0d
Total (in G.B.)	£1,125	5s	0d

EXTRA (inc. P.T.)

Overdrive	£60	13s	11d
Wire wheels	£38	6s	8d

The Spitfire 3 bumper has been adapted to the new GT6, which also has the pseudo-Rostyle wheel covers. More louvres have been added to the bonnet side to assist cooling air flow.

Below: From the rear, visible changes include the big transverse exhaust silencer and twin outlets. There are chrome-plated air outlets from the passenger compartment in the rear quarters.

Bottom: The facia has been re-styled using TR5-type switches and instrument markings, with a matt wood finish to the facia board. New for 1969 are the face level air flow "eyeballs". The wheel now has padded spokes. Overdrive, fitted on this particular car, is still optionally extra.

car's centre line than before, but not quite in line with the wishbone pivot.

The drive shaft now has to accommodate a little plunge and some bending, so a Triumph 1300-type rubber doughnut and spider has been added as far outboard as possible, really as an "outboard" universal joint. For space reasons it is nevertheless a few inches inboard of the hub carrier, and thus perched out in mid-shaft it looks bulky, but there is apparently no problem with drive shaft critical whirl periods as the maximum rotational speed envisaged is only about 1,800 rpm. The cast bottom wishbone has had to be deeply bowed and the telescopic damper re-positioned to give clearance to the rubber doughnut; the damper's top mounting is now on the bodyshell itself, in the wheel-arch. The original damper mounting is still present, but not redundant. Total wheel camber changes (from full bump to full rebound) are greatly reduced from 21deg. to 7deg. 20min., with almost vertical wheels at normal driving loads.

There are no changes to the brakes or wheels. Wire wheels are optional extras, Dunlop SP41 radial ply tyres standard, and the disc wheels now have the Pseudo-Rostyle wheel trims (complete with false wheel nuts!) first seen on the TR5.

Power output is up from 95 bhp (net) at

TRIUMPH Mk2 GT6

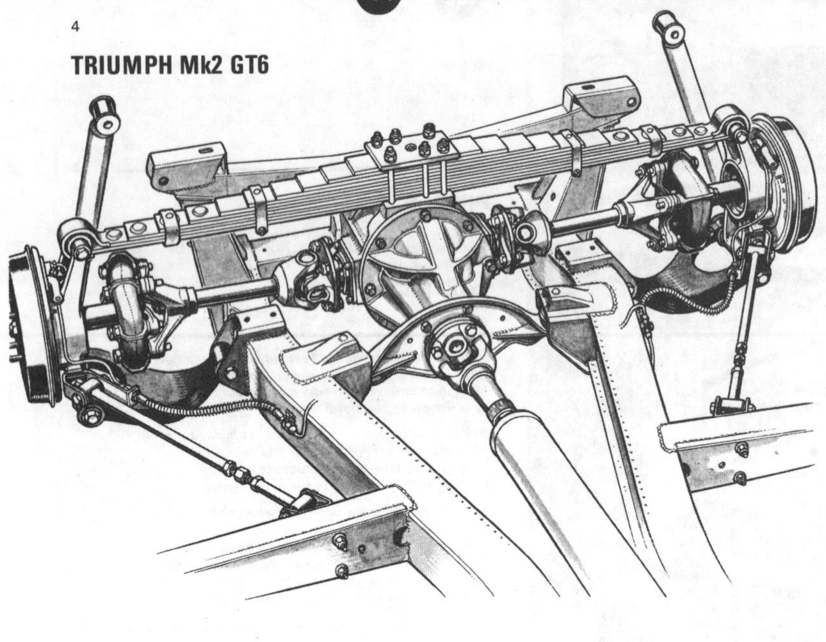

Above: The original GT6 chassis frame has not been chopped about, but new brackets have been added to support the wishbone's inner pivot, and radius arm pivot pick ups have been moved inboard. The slim telescopic dampers are now mounted on the body in the wheel-arch. The transverse leaf spring forms the upper "wishbone" and drive shaft plunge is taken by the 1300-type doughnut. Camber change from bump to rebound has been reduced drastically.

Summary of specification changes

	Original Mk1	New Mk2
1,998c.c. engine	95bhp(net)at5,000 rpm 117lb.ft.at 3,000rpm	104bhp(net)at 5,300rpm 117lb.ft.at 3,000rpm
Final drive ratio	3.27to1 withO/D3.89	3.27to1 withO/D3.27 3.89tospecialorder
Rear suspension	i.r.s., swing axle, transverse leaf spring	i.r.s.bottomwish-boneanduppertrans-verse leaf spring
Heater	Extra, no face level vents	Standard, includes face level vents

Above: Worm's-eye view of the new suspension showing the new bottom wishbones bowed down to clear the big doughnuts, and the damper whose top mounting is on the wheel arch.

5,000 rpm to 104 bhp (net) at 5,300 rpm. This has been done very simply by adopting the TR5's full-width cylinder head, valves and porting (but a different combustion chamber depth for the required com pression ratio) by new exhaust manifolds (taken from the TR250 2½-litre car—the US version of the TR5) and new camshaft timing. The head gives better breathing because of improved port runs, while camshaft timing is changed from the previous 18–58–58–18 to 25–65–65–25 deg. Valve lift is up from 0.312in. to 0.336in., the new cam profiles being exactly the same as the Triumph Spitfire 3. The 1,998 c.c. cylinder block is now completely interchangeable with the TR5's 2,498 c.c. block, this change having come about for the Mk 2 GT6.

There are no transmission changes except that the optional overdrive is now offered with the same (3.27) axle ratio as non-overdrive cars. This gives the very high top gear ratio of 2.62, and a mph/1,000 rpm figure of 25.2 mph. The original 3.89 ratio is still available to special order.

Previously the heater was an extra, but the Mk 2 system is now standard and includes face-level eyeball vents in the facia together with air outlets, rather garishly plated, behind the rear quarter windows. A heated rear window is also standard, so general standards of vision and ventilation are much higher than before.

External styling changes are few but obvious. At the front, the Spitfire "bone-in-teeth" bumper has been fitted, there are Mk 2 badges at front and rear, and in the sides of the swing-forward bonnet are additional louvres to let out the hot air that is a minor problem with this well-packed engine bay. There is a new exhaust system with massive transverse silencer to go with the uprated engine, and this together with twin outlets are obvious from the rear.

Inside the car, the facia is completely new, bearing some relationship to the TR5 design. The wooden facia board now has a matt finish, "safety" rocker switches are fitted, and the steering wheel spokes have padded covers. There is a bit more headroom due to re-contoured seats.

With all these improvements, there has had to be a price increase, from £1,024 to £1,125, but this is reasonable in view of the number of previous extras now standard and the much more advanced specification. Deliveries of the new car begin at once. **G.R.** □

GT6

WHEN THE QUESTION OF TRANSporting the Deputy Ed to Madrid for the Spanish GP arose all sorts of suggestions were put forward ranging from a standard Min to a Rolls-Royce Silver Shadow. The Rolls was unavailable as usual and the Deputy Ed came over all queer when the idea of a Min was mentioned, so after a lot of argy bargy a Triumph GT6 was settled upon.

Everyone on the staff recoiled with horror when I chose STI's six-cylinder coupé but I had the advantage of them—I hadn't driven it! At least I hadn't driven it on the road due to the agility of the Standard-Triumph press department, who had managed to sidestep all my requests for a road test car, but I had driven it at Mallory Park when it was first being introduced to the press. To my surprise I had found then that it handled well on the tight turns, and I had a disgraceful race with fellow scribe Bill Gavin, both of us urged on by other journalists, while the white-faced PR types stood by waiting for the inevitable shunt. It ended when Bill spun at the chicane and I went on to victory. However, I digress. Driving a car on the road can be a vastly different matter from driving on the race-track and my colleagues warned me that all was not well.

The plot was to drive from England to Spain as quickly as possible to save precious time away from the office, so I timed my exit from Lydd to Le Touquet to give me the hours of darkness in which to put as much distance between the GT6 and the English coast as possible. Having measured the wall map with a thumb nail we decided that it was about 800 miles from Le Touquet to Madrid, but our thumb nails must be a bit short because the Triumph's odometer had clicked up 1100 miles by the time I cruised into Madrid exactly 23 hours after leaving the French coast. Omitting an hour for a grotty dinner in Rouen and another hour for a rather strange lunch at Burgos the next day I averaged 52.5mph, which is not slow.

I rather like the idea of a big engine in a small chassis as this usually gives benefits in acceleration, top speed and a lower noise level. Where it can fall down is in the way it affects handling, steering, ride and braking. Alas, this is just where the GT6 suffers, for it is fairly obvious that dropping a two-litre 95bhp six into what is virtually a Triumph Herald chassis is going to be exciting to say the least. The swing axled Herald is not the best handling car on the roads and neither is the GT6, although there are constant rumours that the swing axles are to be replaced by something more modern such as the trailing arms of the 2000/TR5 range. On smooth British roads the car feels quite pleasant, understeering up to reasonably high cornering speeds and power-oversteering on demand, but on the rough roads of France and the even rougher ones of the Spanish N1 to Burgos and Madrid the car was constantly put off line by the bumps and cornering became a very hit or miss affair.

Even worse, the machine would suddenly leap sideways on a bumpy straight for no apparent reason, which was potentially dangerous if one happened to be near the edge of the road at the time for the back wheels would suddenly be bouncing along in the rough. The ride, too, when fully laden is not beyond criticism; the car threw its occupants about something wicked on bad roads and the rear suspension was constantly bottoming, jerking the base of my spine time and again. This bottoming invoked rear wheel steering and a peculiar corkscrew motion.

Although the GT6 has a cruising potential of 100mph or more I seldom let it get much above 80 on the continent as it was just not controllable. The engine is one of the nicest features of the car, remaining smooth and quiet and moving the GT6 away from rest quite smartly. Quite what the exact performance figures of this particular car are we don't know and it's not much help scratching through *Motocar* for they differ by as much as 4.7sec to 90mph and by as much as 6mph in their estimate for the maximum speed of first gear, both of them using 6000rpm and the same axle ratio. Our Giant Test figures (Oct '67) were nearer the quicker ones achieved by the *Moto* bit of *Motocar.* The gearbox has good synchromesh and ratios seem to be well spaced but changes are rather sticky and it invariably refused to engage first, especially when cold. The expensive overdrive which works on third and top was extremely jerky in operation, needing a blip of the throttle or a dip of the clutch to cushion the jolt.

Conditions inside the car can best be described as cramped. The seats are comfortable enough but a 6ft driver is a tight fit, with minimal body room and visibility, not too noticeable on short runs but conducive to the screaming habdabs after 1100 miles. Luggage space is good for a sports car but by the time you have piled in a couple of cases, a typewriter and odd coats and things the vision through the rear window cum door is somewhat reduced.

Despite my disappointment with the GT6 I still think that the idea of a big engine in a small chassis can work. If Triumph can be persuaded to make the rear suspension more sanitary and find a bit more room inside, the GT6 could rival the MGB. On the other hand it may just disappear when the British Levland Motor Corporation gets really amalgamated. And that would be a pity as the potential is there. *MLT*

TR GT 6

$3045
West Coast P.O.E.

Derived from the Spitfire, fastback coupe with engine from the 2000 sedan. Quiet and smooth; performance very good. New rear suspension. Excellent progressive brakes; throttle response very good. Road holding good. Ventilation, improved; visibility limited, and interior cramped. Finish, very good. Price, quite reasonable. Manufactured by Standard-Triumph International Ltd., Coventry, Warwickshire, England.

Triumph calls it the GT 6+. And the things about it that are '+' are the very things that we found to be '–' last time we tested the car.

There's an immense satisfaction in reporting candidly on a vehicle and then coming back to it the following model year to find that the very things we considered to be shortcomings haven't merely been corrected — they've been transformed into strong points.

That's exactly what's happened in the little Triumph fastback. The worst faults of the GT 6 have become some of the greatest virtues of the GT 6+.

From the time Triumph lifted the 1998 cc engine out of its 2000 sedan, dropped it into the Spitfire chassis and covered the whole thing with a slick fastback body, it's been something of a baby supercar, a miniaturized *gran turismo*, with one whale of a lot going for it.

And a lot against it, too.

The GT 6 looked good and it went well. Its styling appeal, basically unchanged since the car was introduced, is obvious. Its performance, exemplified by a standing quarter-mile in the low 18's, is the best England has ever offered in a $3,000 sports car — the best, indeed, that England has ever offered in *any* $3,000 car.

But its handling . . .

The original GT 6 had the Spitfire's rear suspension, swing axles with transverse leaf springs. This, in turn, was a legacy from the Triumph Herald, a nimble little economy car that never really had enough horsepower to create any handling difficulties.

But, by the time that same suspension had come from the Herald through the Spitfire, it had — uh, as our English cousins might say — outlived its usefulness. And, by the time it had reached the GT 6, there were more than a few colonials who were ready to call at the Triumph factory to re-enact the battle of Yorktown. Indeed, the car gained an early reputation among even the most anglophilic sports car enthusiasts as 'George III's revenge.'

But, now, there is liberty and justice for all and a new way to pursue happiness. No matter what Valley Forge they may have come from, the components that form the rear suspension of the new GT 6+ *work*.

Triumph has created a totally new and fully independent rear suspension for this remarkable little vehicle.

Two Stromberg CD carburetors fuel the GT6 engine and give good performance while keeping within the exhaust emission regulations.

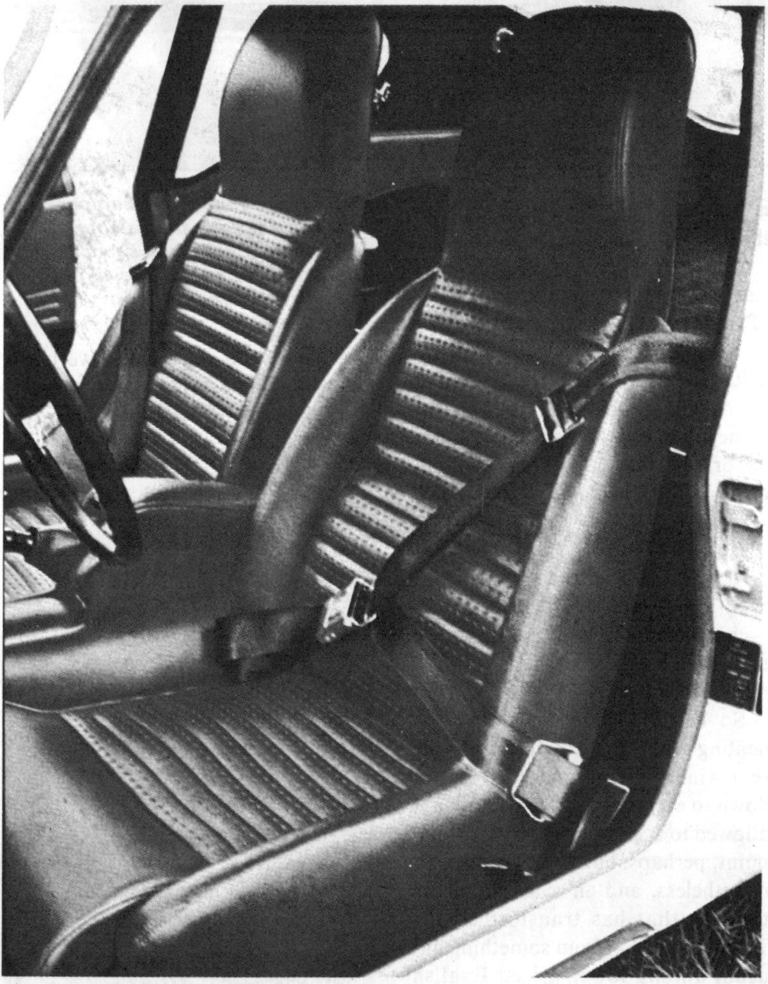

Ventilation, correct contours, and built-in headrests make the GT6 seats both comfortable and safe.

The axle shafts are now fully articulated — a ponderous and Britannic way of saying there's a 'U' joint at each end of each axle shaft and that, therefore, there's no longer the excessively Teutonic camber change that occurs with swing axles.

We wish we had a Celtic gift for words to describe the difference in the way the car behaves on the road. No longer does the GT 6 squat severely under the hard acceleration that is its forte. True, the rear end still settles a bit when the enthusiastic driver tries for one of those 18-second quarters. But bite is solid. And the front end and hood don't leap into one's line of vision. The car simply settles down to the business of going quickly.

This improvement in straightaway stability alone was responsible for cutting a good quarter of a second off the standing quarter-mile time we recorded with our last GT 6. From 18.2 seconds last time, we were able to pull the car down to 17.95.

But it took work. We'll still call it in the low 18's as far as the average driver, with a showroom-fresh GT 6+, is concerned.

If the car proved to be better handling and performing on the straight, it was absolutely sensational when we took it down a few of our favorite winding roads. As if the Corvair, Volkswagen and, of course, the Porsche hadn't already made the point clear, the difference between swing axles and fully articulated rear suspension is . . .

Do you know Gladys on 'Laugh In?' And do you know Raquel Welch?

From an unpredictable, oversteering handler, the GT 6 in its new '+' form has evolved into a cornering demon. It not only performs but it now handles too. There is still a measure of twitchiness when the car is thrown through a succession of turns with varying radii. One does have to bring it out of one turn with some thought of setting it up for the

The GT6 fastback body shape is derived from works Spitfires that ran in the Le Mans 24 Hour race three or four years ago.

Though the high front bumper runs right across the radiator air intake, it does not seem to affect the cooling of the two-liter, six cylinder engine.

next. But, in skilled hands, it calls for no more than a thought.

Gone is the sudden oversteer characteristic of vehicles with swing axles. The GT 6+ corners with an initial understeer that becomes oversteer only with application of throttle. The rear end will come about not spontaneously but only on command performance. And, without the severe camber change swing axles produce, what oversteer there is available is entirely an accelerative matter. Therefore, combined with the vehicle's inherent understeer, it is a fully controllable quality.

We've laid stress on the redesigned rear springing of the GT 6+ because it's probably the most important of the car's improvements over the old GT 6. But there are other changes which deserve praise, too.

In this hardtop era, it's difficult to imagine a car that looks better with the windows up than with them down. But, in this high-performance era, it's difficult to believe that anyone would offer as lively and, therefore, as noisy a machine as the GT 6 without providing some consideration for driver fatigue.

In brief, drive a GT 6 with the windows down and breathe freely but know the roar of the wind. And the 95-horsepower, six-cylinder engine.

Or drive a GT 6+ with the windows up and breathe freely. And know what that last sigalert was in time to find an off-ramp and avoid it. Hear what is within the car and throw away those wind-cheating snoods that went out with the MG TD.

The TR 6+ has a new flow-through ventilation system, the first of its kind on any English sports car that will turn an 18-second quarter and sells for less than $3,000. On warm days, one no longer has to lower the glass for comfort. Fresh air is automatically circulated through the interior.

Then there was one fault that never caught our notice, though it's the sort of thing that engages the attention of safety engineers. Fastback styling results in a long, flat rear window. And, in inclement weather, that can become a liability. A great lot of snow, ice or whatever collects on the window and becomes most bothersome to remove.

Rolls-Royce, the Ford Thunderbird and the Pontiac Grand Prix, even VW — none of them fastbacks — have come up with an answer to this difficulty ahead of the U.S. Govt. requirement.

So has the Triumph GT 6+. A heating grid is literally fused into the rear window. Cold matter, from ice down to everyday frost, is simply not allowed to accumulate there. A small point, perhaps, but an improvement, nonetheless, and an example of the thought that has transformed the Triumph GT 6+ from something unusual among lower-priced English sports cars into something unique — into very nearly a poor man's XK E.

The styling and performance that have gained the Triumph GT 6 are still there. But, to them, have been added qualities of handling, comfort and convenience that we consider very '+' indeed. ♠

TR GT-6

Data in Brief

DIMENSIONS

Overall length (in.)	147.0
Width (in.)	57.0
Height (in.)	47.0
Wheelbase (in.)	83.0
Track front (in.)	49.0
Track rear (in.)	49.0
Turning diameter (ft.)	25.3
Fuel tank capacity (gal.)	11.7

WEIGHT, TIRES, BRAKES

Weight (lbs.)	1792
Tires	radial ply, 155 SR 13
Brakes, front	disc
Brakes, rear	drum

ENGINE

Type	6 cylinder
Displacement (cu. in.)	122
Horsepower	95

SUSPENSION

Front	independent coil springs
Rear	leaf spring

MORE STING IN THE ENGINE. MORE CLING IN THE TAIL. NEW TRIUMPH GT6 Mk 2.

The Triumph GT6 was a winning streak from the word go. 'Accelerates with a serene ferocity', *Motor* said.

Now the Mk2, with 104 bhp under the bonnet, puts up even whooshier performance. In $7\frac{1}{2}$ seconds you're up to 50.

In top, the figures go like this: 40-60 in 8, 50-70 in 9. But still, the 2-litre engine and its six big cylinders stay smooth. Top speed 110.

Every inch is racy. With its tailored-for-speed bodyline and new raised bumpers it looks rarin' to go standing still.

Every inch reliable. The GT6 Mk2 takes tight corners and fast bends–without wagging its tail. **New wishbone independent rear suspension gives road-hold and cornering power to match the upped performance.**

And in other tight corners, it's worth knowing that the Mk2's turning circle is seven feet smaller than its nearest competitor's.

And underlying all the glamour, there's Triumph engineering. *Safety* engineering. For instance, that streamlined lightweight shape hides the heavyweight strength of a chassis with a double backbone of channelled steel girders. Radial ply tyres are now standard for extra tyre life and added safety. The big-grip, 9.7 in. diam. front wheel discs stop you on a sixpence. The adjustable steering wheel telescopes on serious impact. The rear window is heated. And the doors have anti-burst locks.

The price. £1,125 5s. ex-works, including heater and P.T.

Standard-Triumph Sales Ltd
Coventry. Tel: OCO 3-75511

BRITISH LEYLAND TRIUMPH

Continued from page 52

and maximum revs. So really a set of 45s is needed to adorn the Cox-designed inlet manifolds. Peter also made the free-flow exhaust manifolds which feed into twin $1\frac{1}{4}$ inch unsilenced pipes.

Much of the success of any competition power unit must lie in the quality of the cylinder head modifications, which in this case were done by Ernie Dugmore, who runs a small head-doctoring and engine-rebuilding shop at Radford, Coventry. Like all head-modding geniuses his secrets for perfection are in his brain and expert hands alone, so even Peter knows only that the GT 6 head is 'flowed and generally carved about.' But from past results from the Dugmore-modified Spitfire heads there's little doubt that the mods will be the ultimate. The head started life as the latest production GT 6/2000 'wide head' with the advantages of longer and more efficient porting. Improvements Ernie could disclose were the fitting of larger inlet valves, nimonic like the standard size exhaust valves. Compression ratio is 12:1.

Peter made dural pushrods, lightened the base tappets, fitted solid steel instead of cast iron rockers, and stronger valve springs. Activating all this valve gear is the wildest cam useable in the 2 litre unit, made specially to Peter's own requirements with inlets opening and exhausts closing at 60 degrees and inlets closing and exhausts opening at 80 degrees.

Block and crank are standard Triumph items, but the flywheel bears no resemblance to that in an ordinary GT6. Machined from aluminium it weighs next to nothing.

Meanwhile, back in his shed Ernie was boring out the block to insert some nice, shiny, chrome liners, too hard for the very short, solid-skirt Hepolite Racing pistons to digest, but still showing 1,998 ccs on any over-enthusiastic scrutineer's measuring devices. The Cox kitchen table took a hammering while Peter polished up the standard con rods, and then rods, crank, pistons, flywheel and front pulley gathered together, Ernie balanced the lot, before the crank was mated to the block using standard bearing shells.

Moving the engine rearwards away from the front cross-member enabled the sump to be extended forwards, giving an extra one-gallon capacity of that necessary lubricating fluid. At the same time the oil pick-up pipe from the standard pump was moved to the centre of the new-shape sump and baffles made up round it to stop surge. A 10-row oil cooler mounted in front of the radiator keeps the two-gallons of lubricant at a safe temperature. Water cooling is taken care of by a Triumph 1300 radiator mounted at an angle instead of vertically.

Surprisingly in a car built to this sort of specification, the gearbox is bog-standard GT6, as is the Laycock diaphragm clutch, though lined with Borg and Beck competition material. Only trouble so far has been a broken mainshaft in practise for the Silverstone AP meeting, which meant bolting in a spare box before the Race.

Lucas Competitions Department performed their usual wiring tricks and sorted out a competition distributor to suit the engine.

With the car ready to Race, Peter was running desperately short of money to put it anywhere near a circuit, until at the last minute the Gold Seal Car Co stepped in once again to fill the depleted coffers.

Now this article is meant to be about a GT6, as we've said several hundred times, but it's also about a Spitfire, believe it or not. That sounds like nonsense, but the fact is that part way through the season the car did become a Spitfire in GT6 clothing, when Peter decided to enter the Barcelona 12 Hour Race with one of his old 1000 cc Spit Racing mills dropped in place of the Six. This involved a complete rebuild to move the transmission forward again, alter all the suspension to allow for the weight differences in the units, and 1001 other jobs. This vast reconstruction job being completed, the organisers decided to cancel the Race and all Peter's work was wasted. Out came the 1000 engine and back went all the GT bits without the Spitfire having turned a wheel. A large lump of the season thrown away for nothing! What did we say about bad luck?

Plans for next season are in the balance at the moment. The Cox brain had been working on an idea for a one-off litre monocoque GT car and the sale of the GT6. Such a project would need too much bread, so that's another interesting machine gone by the board. If anybody came along with the right amount of cash he would still sell the GT6 and offer to look after it for the season for the new owner. Failing that he'll continue to Race it, but money's still very short, and as the Gold Seal tie-up has ended he needs a sponsor with hard cash to offer. Then he could settle down and try to realise the full potential of this remarkable car, a tribute to his own engineering genius. That's what we'd like to see happen! **JCR**

GORDON CHITTENDEN PHOTOS

TRIUMPH GT6+

Better handling, quicker, more comfortable

THE NEW VERSION of Triumph's GT6 coupe will be known as Mk II in the British and European markets, but in the U.S. it will be called GT6+ because of certain differences required to meet our safety and exhaust emission regulations. A revised cylinder head gives 11 bhp more than before, but since the new rating is given as 95 bhp at 4700 rpm for the American version, vs. an advertised 95 at 5000 last year, we were confused. Triumph's answer is that the previous figure was a *gross* rating, for the bare engine, whereas the new one is a *net* rating, for the engine complete with accessories and exhaust system. By the net rating the previous engine developed only 84 bhp at 4600 rpm.

The other mechanical change—a major one indeed—is a new rear suspension system. Gone are the swing axles, in their place a system more like the Corvette's than any other we can think of. These consist of double-jointed halfshafts leading to uprights located by lower A-arms at the bottom and at the top by the same transverse leaf spring that was used only as the springing medium in the earlier car. Adjustable trailing arms complete the rear wheel location, and it is remarkable how little chassis modification was necessary to incorporate the new system. The roll center has been lowered from 12.8 in. above ground to 6.3 and camber change is also cut in half. Roll stiffness at the rear is up about 15 percent, mainly because the spring rate is up from 84 lb/in. to 93 (at the wheel); but lateral weight transfer is actually reduced at the rear, thanks to the lower roll center.

On the outside the body shape is the same but the front bumper has been moved up to the same position it occupied on the similar-bodied Spitfire Mk 3, hiding the former park-turn light position and requiring new light units under the bumper. Wire wheels, formerly standard, have been replaced by the cheaper but functionally superior disc wheels, in turn covered up by gimmicky "mag-styled" covers. Real alloy wheels are available as an extra.

Triumph must have bought a new louvering machine: each side of the GT6+ carries two sets of louvers. Both are functional: the ones in the front side panel vent the engine compartment, the chromed ones in the rear quarters serving an improved ventilation system which includes fresh-air nozzles in the dashboard. An electrically heated rear window, with a warning light so that it won't be left on unintentionally, also adds to weather provisions and is standard.

TRIUMPH GT6+

For the safety regulations all the toggle switches have been eliminated from the dash panel and the new heater controls have been recessed into a nice dent in the center. The highly polished wood panel has been replaced by a flat-finished, but still genuine, piece of wood. Naturally, the collapsible-on-impact steering and high-rise "throne" seats meet the 1969 regulations. The new seats improve the Comfort Index ratings and restrict vision to the rear somewhat—these are used only on the U.S. GT6+.

Despite confusion on the part of both ourselves and the Triumph people concerning the torque curve of the +'s engine, the car does outperform its predecessor by a margin that's detectable in daily driving as well as in the performance figures. It gets through the quarter-mile 0.8-sec faster than the earlier version and does a couple mph better in top speed. Fuel economy remains just the same, 24.2 mpg overall. The 6-cyl 2-liter ohv engine of the GT6+ is still the most distinctive part of the car's personality, being smooth in the extreme and offering a throaty 6-cyl purr that isn't available in any other car of its class. This engine has bags of torque, and with the 3.89:1 final drive that comes with overdrive (it's 3.27 without) it will pull strongly from *10 mph* in 4th! The OD, of course, gives it a good cruising gear. Valve float sets in at 5700 rpm, limiting top speed to 98 mph in 4th; overdrive is good for 108 mph at just over 5000 rpm. True to its nationality, the GT6+ runs on quite badly when turned off.

The shift linkage still has that strange, curved H-pattern, making 3-4 shifts a hooker to the left. Our test car had a weak 2nd-gear synchro we hadn't noted in the earlier GT6, and its clutch engaged almost at the top of its pedal travel. Heel-and-toeing isn't easy with the high throttle pedal and

Rocker switches and flat-finish wood meet safety standards.

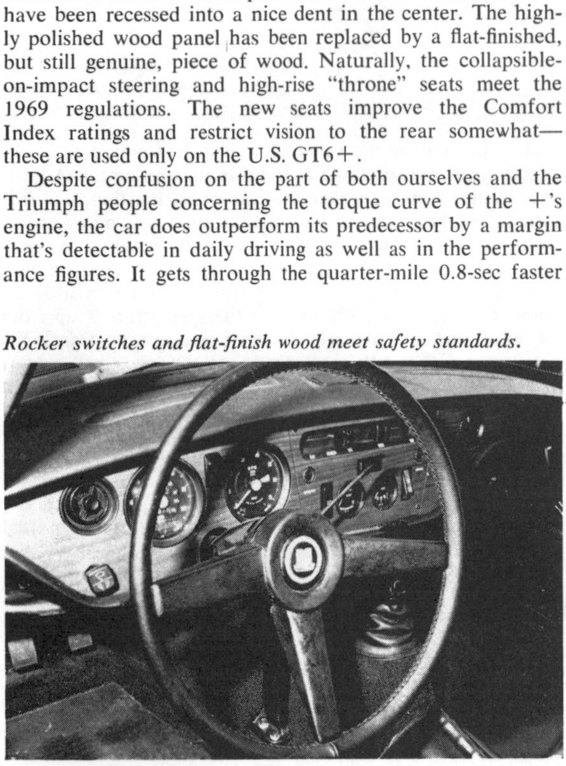

Dome light comes on when tailgate-window is lifted.

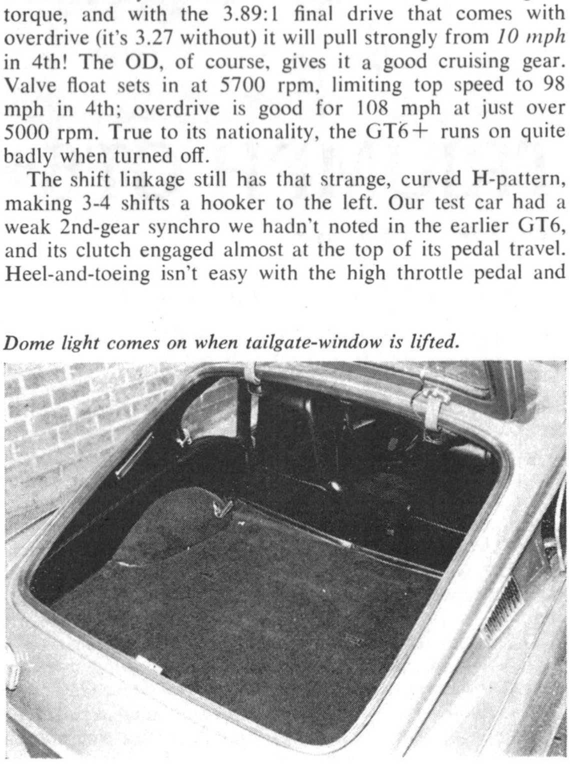

a steering column flange interferes with a large throttle foot. In spite of one Lotus Elan-like rubber coupling in each rear axle shaft, however, it is easy to drive the + smoothly.

With the new rear suspension Triumph recommends equal tire pressures (24 psi) front and rear; we found the tires too soft at this setting, seemingly accentuating the large degree of understeer already present—so we tried 30 psi all around and liked the more solid handling feel. The previous car, with a negative rear camber setting, handled remarkably well for having swing axles, but if one lifted the throttle foot in a hard corner there was little chance of recovering control before a tail-first spinout. With the new articulated suspension, the attitude change when lifting off is still toward oversteer, but now the transition is smooth and easily controllable. As with the previous setup, there is a considerable amount of popping and clunking from the rear, apparently the result of attaching a multileaf spring to a place where its noises are easily transmitted inside the car.

Steering is light and precise, reflecting the moderate front end weight (even though the car is front-heavy), the radial tires and rack-and-pinion gearing. Four-and-a-half turns lock-to-lock sounds like a lot, but it's the incredibly tight turning circle (25.2 ft), not slow steering, that's responsible.

On smooth but undulating roads the GT6+ travels with rather abrupt motions; on rough surfaces it tracks well with a moderate amount of squeaking and rattling. With the firmer rear suspension it's a bit harder-riding overall than the early GT6, even with 24 psi all around.

The GT6+ is a low car, restricted in headroom and not easy to get into or out of, and its high steering wheel can be a problem for extremely short drivers though the column's reach can be adjusted with wrenches. There are entirely too many control stalks on the steering column when overdrive is fitted, causing us—as we have said before—to dim the overdrive and shift the bright lights. Flow-through ventilation and a good supply of air from dash and underdash nozzles mean lots of air through the interior, but not enough to offset a large amount of cockpit heat if it's warmer than 70 degrees fahrenheit.

Interior finish is pleasant and luxurious, with the exception of ill-fitting door upholstery. There is no locking bin inside the car, though some things could be hidden out of sight beneath the rear compartment floor.

In the face of modifications and additional equipment required by Federal law, plus definite product improvements, Triumph has kept the GT6 price stable by substituting the disc wheels. And though we may find fault with things like the messy body details—the shape is good but there are many gimcracks and evidences of production economies—there is no doubt at all of the great value in this car. Where else can you get a 6-cyl, 100+-mph coupe with a proper chassis, good finish and jazzy looks for $3000? Nowhere *we* know of.

Engine accessibility is one of the GT6's strong points.

TRIUMPH GT6+

ROAD TEST RESULTS

PRICE

List price	$3045
Price as tested	$3380
(incl. AM radio & OD)	

ENGINE & DRIVE TRAIN

Engine	6 cyl inline, ohv
Bore x stroke, mm	74.7 x 76.0
Displacement, cc/cu in.	1998/122
Compression ratio	9.25:1
Bhp @ rpm	95 @ 4700
Equivalent mph	100
Torque @ rpm, lb-ft	117 @ 3400
Equivalent mph	72
Transmission type	4-speed manual with OD
Gear ratios, OD (0.802)	3.12:1
4th (1.00)	3.89:1
3rd (1.25)	4.86:1
2nd (1.78)	6.92:1
1st (2.65)	10.30:1
Final drive ratio	3.89:1

GENERAL

Curb weight, lb	1975
Weight distribution (with driver), front/rear, %	56/44
Wheelbase, in	83.0
Track, front/rear	49.0/49.0
Overall length	147.0
Width	57.0
Height	47.0
Steering type	rack & pinion
Turns, lock-to-lock	4.5
Brakes	9.7-in. disc front, 8.0 x 1.25-in. drum rear, 260 sq. in. swept area

ACCOMMODATION

Seating capacity, persons	2
Seat width	2 x 15.5
Head room	38.0
Seat back adjustment, degrees	none
Driver comfort rating (scale of 100):	
For driver 69 in. tall	80
For driver 72 in. tall	75
For driver 75 in. tall	55

PERFORMANCE

Top speed, OD (5000 rpm), mph	109
Acceleration, time to distance, sec:	
0–100 ft	2.8
0–250 ft	6.2
0–500 ft	9.6
0–750 ft	12.5
0–1000 ft	15.0
0–1320 ft (¼ mile)	18.0
Speed at end, mph	77
Time to speed, sec:	
0–30 mph	4.0
0–40 mph	5.9
0–50 mph	8.0
0–60 mph	11.0
0–80 mph	19.5
0–100 mph	37.0

BRAKE TESTS

Panic stop from 80 mph:	
Deceleration rate, % g	87
Control	good
Fade test: percent of increase in pedal effort required to maintain 50%-g deceleration rate in six stops from 60 mph	20
Overall brake rating	very good

SPEEDOMETER ERROR

30 mph indicated	actual 28.8
50 mph	48.7
70 mph	68.3

CALCULATED DATA

Lb/hp (test weight)	24.2
Cu ft/ton mi	109
Mph/1000 rpm (OD)	20.9
Engine revs/mi	2870
Engine speed @ 70 mph, rpm	3320
Piston travel, ft/mi	1430
R&T steering index	1.13
R&T wear index	41.1
Brake swept area, sq in/ton	226

FUEL

Type fuel required	premium
Fuel tank size, gal	11.7
Normal consumption, mpg	23.9

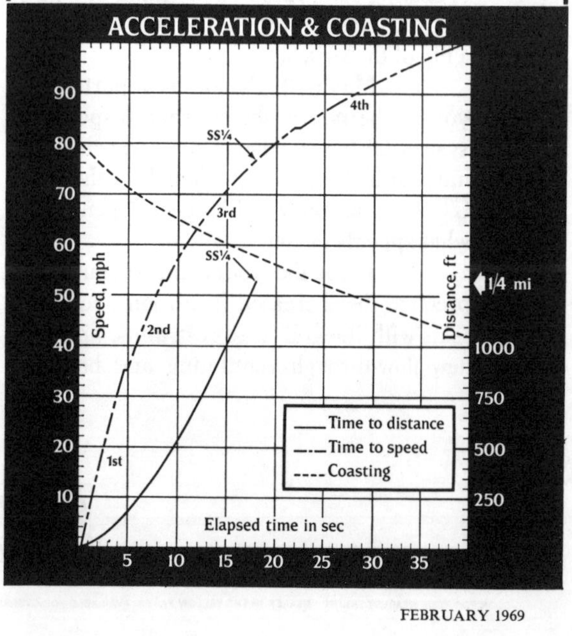

The fasterback with the fog-free window.

The rear window on the new '69 Triumph GT 6+ is heated electronically. A heating grid, fused on the inner surface, frees the glass of fog, ice and snow. All with a flip of the switch.

Maybe that's not as important to you as the new wishbone rear suspension system. Or the fact that all four wheels have independent suspension. Which of the new GT 6+ features you like best depends on what appeals to you.

If you like your comfort you'll appreciate the increased headroom, contour seats with integral head restraints, and the new flow-through ventilating and heating system with face and floor-directed vents.

If style is your thing the GT 6+ has a new self-sealing magnetic gas cap, mag-style wheel covers and white walls—all as standard equipment.

But not everything's new on the new GT 6+. It's still got a six cylinder high-torque engine under the hood. And it still comes with four forward speeds, all-synchromesh gear box, disc brakes up front, and rack-and-pinion steering.

So whether it's the heated rear window or the other 1969 features, there's a lot in the new GT 6+ worth looking into. At your Triumph dealer.

BRITISH LEYLAND TRIUMPH

'69 GT6+

LOOK FOR YOUR NEAREST TRIUMPH DEALER IN THE YELLOW PAGES. AVAILABLE FOR OVERSEAS DELIVERY. BRITISH LEYLAND MOTORS INC., 600 WILLOW TREE ROAD, LEONIA, N.J. 07605.

MOTOR TESTED

Much improved

New rear suspension greatly improves roadholding and handling; good performance and excellent economy made still better by changes; engine rather noisy at high revs.

N ALMOST every way the Triumph GT6 was favourably eceived by us in our original road test back in October 1966. nly one adverse comment of any consequence marred the eneral approval, and this referred to a fault in the handling: "If ou lift off or, worse still, brake while cornering really hard, nere is a sudden and quite vicious transition to strong oversteer vhich must be smartly corrected with opposite lock." At the time ome of us felt that this defect made the car unworthy to be rouped with the other medium-cost British production sports ars which have been such a success since the war and con- ributed greatly to our export sales.

But no one need now black-ball the GT6 in its Mk.2 form, for nodifications to its rear suspension have transformed the hand- ng and roadholding. Its performance, always good, has been nproved still further by a small power increase giving a naximum speed of nearly 110 m.p.h. with very good accelera-tion, particularly in top, and excellent fuel economy. Worth-while minor changes include a facia improved in layout and appearance, thinner seat cushions to increase headroom, and a much better heating system with proper fresh-air ventilation.

To most people the term "GT" implies a 2+2 seating arrangement and in this respect the car does not live up to its name in the version tested since the rear platform is for luggage only. But it is nevertheless large—children are quite happy to lie on it for short journeys—and easily loaded through the lift-up rear door; rivals with occasional rear seats such as the MG B/C GT and Reliant Scimitar are all more expensive, some of them considerably so. However, an occasional rear seat for the GT6 became available as an option with the announcement of the Mk.2 car at an extra cost of £19 11s. 8d.

RICE: £879 plus £270 17s 6d purchase tax equals £1,149 17s 6d. verdrive £62 0s 3d with purchase tax, total as tested £1,211 7s 9d.

NSURANCE: AOA Group Rating, 6; Lloyd's 6.

Performance and economy

Calculation shows that the power increase from 95 (net) b.h.p. at 5,000 r.p.m. to 104 b.h.p. at 5,300 r.p.m. ought to raise maximum speed from the 107.8 m.p.h. of our "Mk. 1" road test car to a little more than 111 m.p.h., making due allowance for

Triumph GT6 Mk.2 *continued*

tyre drag. In fact the car lapped MIRA's banked track at 109.1 m.p.h. in overdrive top but at this speed there is a good deal of tyre scrub and, of course, as the maximum rises above the much lower "hands off" level, the drag increases rapidly. So there is every reason to suppose that the GT6 will do its 111 m.p.h. or more given a chance on a flat motorway. Much more apparent and useful is the improvement in acceleration, the Mk. II version being more than a second quicker to 60 m.p.h. from a standstill, and progressively better as speed rises further. This increased performance has been achieved at no more than a small sacrifice in bottom-end torque, perhaps due to the increased valve overlap, for the engine will pull smoothly and quite willingly from 20 m.p.h. in overdrive top (950 r.p.m.), although there is a noticeable quickening in tempo above 2,500 r.p.m. Up to about 4,500 r.p.m. the engine is both smooth and quiet, but above this speed it begins to sound fussy, while above 5,000 r.p.m. it is even a little rough for a "six". The unit was a little sensitive to the choke setting for cold starts, and the slow running was less

Performance

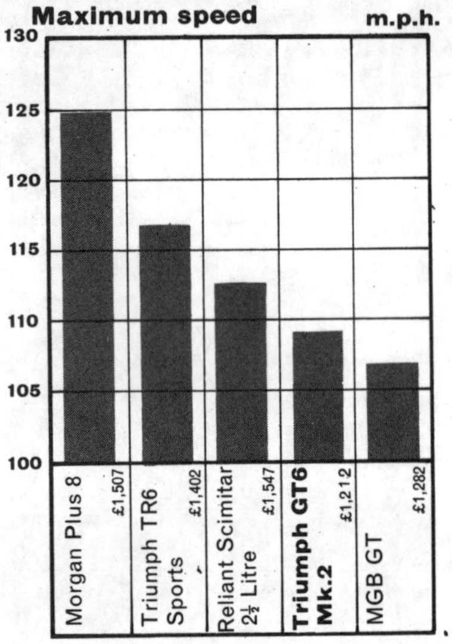

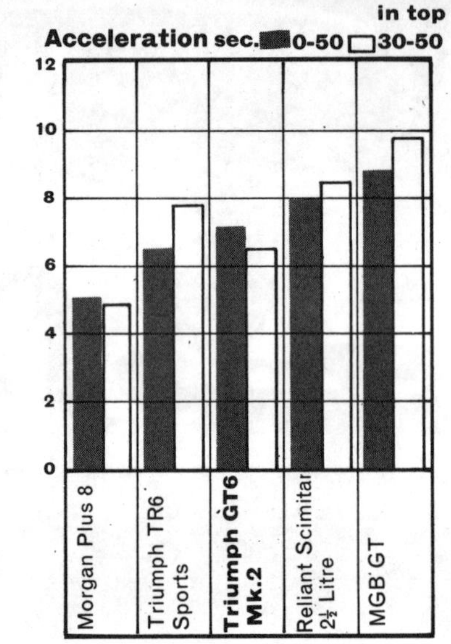

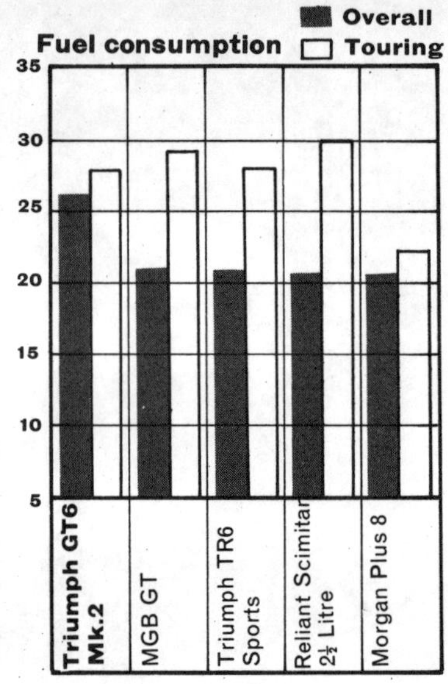

Performance tests carried out by *Motor's* staff at the Motor Industry Research Association proving ground, Lindley.

Test Data: World copyright reserved; no unauthorised reproduction in whole or in part.

Conditions

Weather: Cool and breezy
Temperature: 52°F
Barometer: 29.8in. Hg.
Surface: Dry concrete and asphalt
Fuel: Super Premium 101 octane (RM) 5 star rating

Maximum Speeds

		m.p.h.	k.p.h.
Mean lap banked circuit		109.1	175.5
Best one-way ¼-mile		109.6	176.1
Direct top gear	at 6,000 r.p.m.		162
O/d 3rd gear	at 6,000 r.p.m.	101	130
3rd gear	at 6,000 r.p.m.	81	92
2nd gear	at 6,000 r.p.m.	57	61
1st gear	at 6,000 r.p.m.	38	

"Maxmile" speed: (Timed quarter mile after 1 mile accelerating from rest)

Mean	109.1
Best	109.6

Acceleration Times

m.p.h.	sec.
0–30	3.4
0–40	4.9
0–50	7.2
0–60	9.4
0–70	12.7
0–80	16.9
0–90	22.4
0–100	30.2
Standing quarter mile	17.4

m.p.h.	O/d Top sec.	Top sec.	3rd sec.
10–30	—	—	5.6
20–40	9.2	6.6	5.3
30–50	9.3	6.5	5.1
40–60	8.9	6.6	5.2
50–70	9.2	6.9	5.8
60–80	11.1	7.8	6.7
70–90	16.3	9.7	—

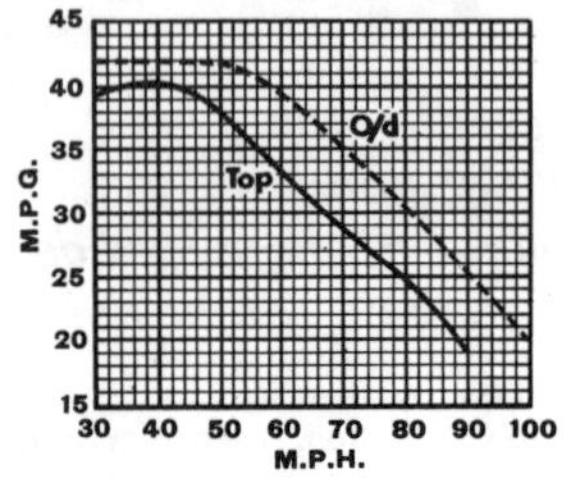

Fuel Consumption

Touring (consumption midway between 30 m.p.h. and maximum less 5% allowance for acceleration) 27.9 m.p.g.
Overall . . . 26.1 m.p.g. (=10.8 litres/100km)
Total test distance . . . 1,570 miles

Brakes

Pedal pressure, deceleration and equivalent stopping distance from 30 m.p.h.

lb.	g.	ft.
25	0.39	77
50	0.60	50
75	0.92	32½
90	0.97	31
Handbrake		

Fade Test

20 stops at ½g. deceleration at 1 min. intervals from a speed midway between 40 m.p.h. and maximum speed (=74.6 m.p.h.)

	lb.
Pedal force at beginning	40
Pedal force at 10th stop	40
Pedal force at 20th stop	40

Steering

Turning circle between kerbs: ft.
Left . . . 23½
Right . . . 23½
Turns of steering wheel from lock to lock . . 4.3
Steering wheel deflection for 50ft. diameter circle . . . 0.9 turns

Clutch

Free pedal movement . . . =½ in.
Additional movement to disengage clutch completely . . . =3 in.
Maximum pedal load . . . =25 lb.

Speedometer

Indicated	10	20	30	40	50	60	70
True	10	19	28½	38	48	58	69
Indicated	80	90	100				
True	80	90	100½				

Distance recorder 1.2% fast

Weight

Kerb weight (unladen with fuel for approximately 50 miles) . . . 17.75 cwt.
Front/rear distribution . . . 56/44
Weight laden as tested . . . 21.5 cwt.

Parkability

Gap needed to clear 6ft. wide obstruction in front 4' 5"

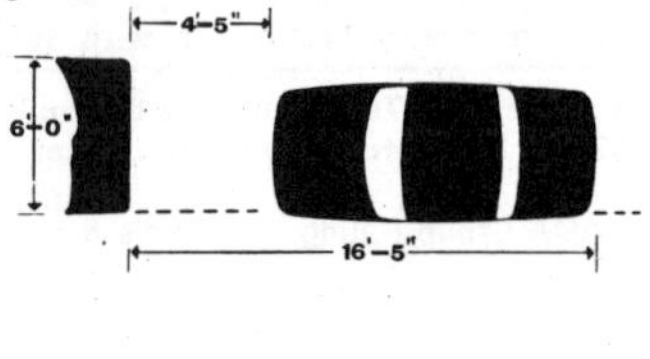

As on the Spitfire, the front bumper has been raised to meet American safety regulations.

Seats have been thinned down to increase headroom but they are still comfortable. Leather-covered steering wheel has a thinner rim than before.

High-speed mobile greenhouse: one of the many uses for the rear luggage space. Note extractor louvres in rear quarters.

Triumph GT6 Mk. 2 *continued*

even than for our earlier car, again perhaps because of the increased valve overlap.

Despite a drop in touring fuel consumption from 31.4 m.p.g. to 27.9 m.p.g., the overall consumption remains virtually unchanged at 26.1 m.p.g. With the 27 m.p.g. which a private owner might be expected to get, the overall range from the 9¾ gallon tank would be about 260 miles—not quite enough for a car that aspires to the GT label.

Transmission

Normally a 3.89:1 final drive ratio goes with the optional overdrive as on our test car, but a 3.27:1 ratio—standard without overdrive—is available to special order to give more restful high-speed cruising. As the car gets to high speeds very quickly with the standard final drive unit, and as the engine begins to sound active in the upper rev hand, we found the car a little fussy in direct top on fast secondary roads when we regarded overdrive top as a usable fifth gear, and not just a motorway cruising ratio.

With more power than before and a lowish bottom gear it is easy to run out of revs in this ratio, though it allowed an easy restart on a 1-in-3 slope. Second and third, however, are well spaced. The change itself was a little stiff—especially into reverse—and the synchromesh on second could be beaten on fast changes. But the clutch and throttle were progressive in action and smooth changes were not difficult to achieve. The indirect gears whined loudly.

Handling and brakes

Undoubtedly the modified rear suspension is the most important feature of the Mk. 2 car. Gone are the Herald-derived swing axles of the older car; in their place is a double wishbone system with the ends of a modified transverse leaf spring acting as the upper wishbones and with forked transverse links and radius rods below. The new layout has banished the tendency to vicious oversteer that we had cause to criticize sharply in our last test and makes the GT6 a much safer and more predictable car. Nevertheless, we were a little disappointed in the roadholding which we don't regard as outstanding for an all-independent modern sports car. (Failings of this sort cannot, we feel, be ascribed to a lack of stiffness in the separate chassis structure, for there was no scuttle shake and the car has an excellent one-piece feel.)

continued overleaf

Seven cubic feet of our test boxes could be loaded on to the rear platform without obscuring the driver's backward view.

Cornering is a much more happy state of affairs with the revised rear suspension arrangements.

The GT6 is probably one of the happiest blends of open sports car and closed top, looking very much "as a whole".

The slight static negative camber of the rear wheels is not apparent in this view, but the wires of the heating element of the backlight are clearly visible. Also shown is the rear "GT6 Mk.2" badge.

Specification

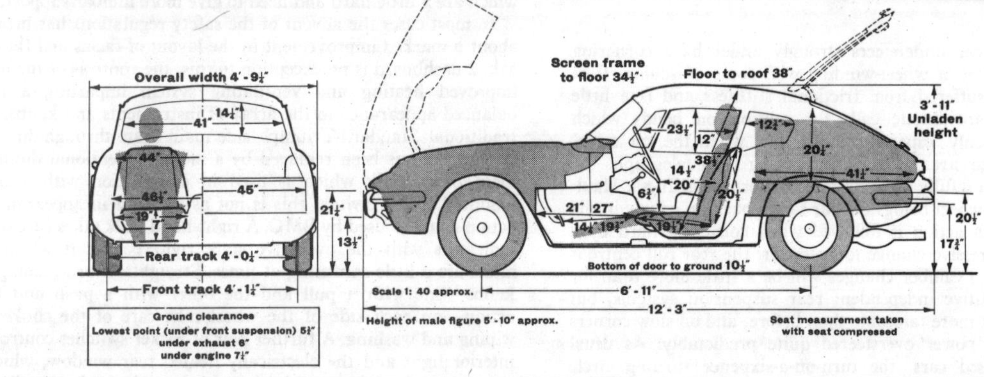

1,998 c.c. 6-cyl engine, rear-wheel drive, all-independent suspension, separate chassis.

Engine

Block material	Cast iron
Head material	Cast iron
Cylinders	6
Cooling system	Water
Bore and stroke	74.7 mm. (2.94 in.) 76 mm. (2.992 in.)
Cubic capacity	1,998 c.c. (122 cu. in.)
Main bearings	4
Valves	Pushrod o.h.v.
Compression ratio	9.25:1
Carburetters	Two Stromberg 1.5CD
Fuel pump	Mechanical
Oil filter	Full flow
Max. power (net)	104 b.h.p. at 5,300 r.p.m.
Max. torque (net)	117 lb.ft. at 3,000 r.p.m.

Transmission

Clutch	8½ in. s.d.p. diaphragm
Internal gearbox ratios	
Top gear	1.00:1
3rd gear	1.25:1
2nd gear	1.78:1
1st Gear	2.65:1
Reverse	3.10:1
Overdrive top	0.80:1
Overdrive third	1.00:1
Synchromesh	All forward ratios
Overdrive type	Laycock de Normanville
Final drive	Hypoid bevel 3.89:1 (3.27:1 without overdrive or as option)
M.P.H. at 1,000 r.p.m. in:—	
O/d top gear	21.2
Top gear	16.9
O/d 3rd gear	16.9
3rd gear	13.5
2nd gear	9.5
1st gear	6.4

Chassis and body

Construction	Separate chassis

Brakes

Type	Discs/drums
Dimensions	Discs 9.7 in. dia.; drums 8 in. dia.
Friction areas:	
Front:	22.2 sq.in. of lining operating on 197 sq.in. of disc
Rear:	38 sq.in. of lining operating on 63 sq.in. of drum

Suspension and steering

Front	Independent by wishbones and coil springs with anti-roll bar
Rear	Independent with transverse upper leaf spring, lower wishbones and radius arms
Shock absorbers:	
Front and Rear	Telescopic
Steering type	Rack and pinion
Tyres	155 SR 13, Dunlop SP68
Wheels	13 in.
Rim size	4½J

Coachwork and equipment

Starting handle	No
Tool kit contents	Two open-end, one box and one plug spanner; one combination tool.
Jack	Scissors type
Jacking points	Under frame at any point
Battery	12 volt negative earth 56 amp hrs capacity
Number of electrical fuses.	3
Headlamps	45/40W sealed beam
Indicators	Self-cancelling flashers
Reversing lamp	No
Screen wipers	Two-speed electric self-parking
Screen washers	Manual
Sun visors	Two
Locks:	
With ignition key	Doors
With other keys	None
Interior heater	Fresh air
Upholstery	PVC
Floor covering	Carpets
Major extras available	Overdrive, wire wheels

Maintenance

Fuel tank capacity	9¾ galls.
Sump	8 pints SAE 20W/50
Gearbox	1½ pints SAE Hypoid 90
Final drive	1 pint SAE Hypoid 90
Steering gear	Grease
Coolant	11 pints (2 drain taps)
Chassis lubrication	Every 6,000 miles to 2 points
Minimum service interval	6,000 miles
Ignition timing	10° b.t.d.c.
Contact breaker gap	0.014-0.016 in.

Sparking plug gap	0.025 in.
Sparking plug type.	N-9Y Champion
Tappet clearances (cold)	Inlet 0.01 in.; Exhaust 0.01 in.
Valve timing:	
inlet opens	25°b.t.d.c.
inlet closes	65°a.b.d.c.
exhaust opens	25°b.b.d.c.
exhaust closes	65°a.t.d.c.
Rear wheel toe-in	$\frac{1}{16}$-$\frac{1}{8}$ in.
Front wheel toe-in	0-$\frac{1}{16}$ in.
Camber angle.	$2° \pm \frac{1}{2}°$ positive, front; $2° \pm \frac{1}{2}°$ negative, rear
Castor angle	$4° \pm \frac{1}{2}°$
King pin inclination	$6\frac{3}{4}° \pm \frac{3}{4}°$
Tyre pressures:	
Front:	20 p.s.i.
Rear	24 p.s.i.

Safety check list

Steering Assembly	
Steering box position	Well back, beneath engine
Steering column collapsible	Yes
Steering wheel boss padded	Yes
Steering wheel dished	Very slightly
Instrument Panel	
Projecting switches	No
Sharp cowls	No
Padding	Above and below facia
Windscreen and Visibility	
Screen type	Zone toughened
Pillars padded	No
Standard driving mirrors	Interior
Interior mirror framed	Yes
Interior mirror collapsible	Yes
Sun visors	Crushable
Seats and Harness	
Attachment to floor	On slides
Do they tip forward?	Yes
Head rest attachment points	No
Back of front seats	Unpadded
Safety Harness	Lap and diagonal
Harness anchors at back	No
Doors	
Projecting latches	Yes
Anti-burst locks	Yes
Child-proof locks	No, only two doors

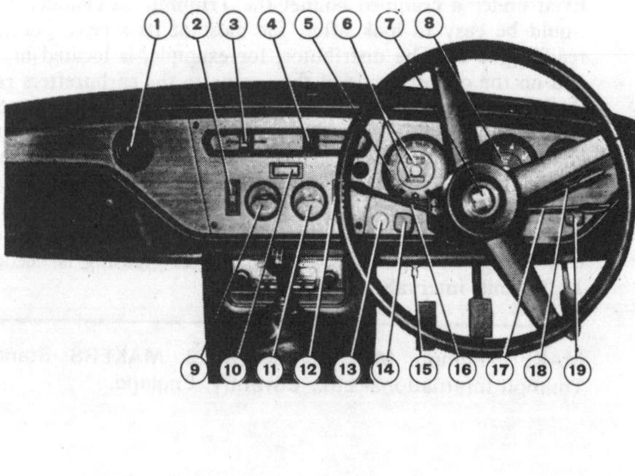

1, fresh air vent. 2, interior light switch. 3, heater distribution and booster lever. 4, heater temperature lever. 5, speedometer. 6, total and trip mileometers. 7, horn button. 8, rev-counter. 9, temperature gauge. 10, backlight heater switch. 11, fuel gauge. 12, lights switch. 13, ignition/starter lock. 14, choke. 15, trip reset. 16, main beam/flasher stalk. 17, indicator stalk. 18, overdrive stalk. 19, wash/wipe control.

Triumph GT6 Mk.2 *continued*

Initially, the car understeers strongly under hard cornering and demands a lot of wheel-winding with some muscular effort as the steering suffered from frictional stiffness and is a little heavy. This characteristic calls for caution on bends which tighten up suddenly, especially in the wet when the car sometimes tends to go straight-on. The steering then telegraphs the event by going usefully light despite its normal stiffness. Hauling hard on the wheel brings the car safely back on course in the dry, though such action is followed when cornering hard by a reasonably progressive change to oversteer: the rear roll centre is still quite high so camber changes will be a little more than for the best competitive independent rear suspension systems, but the change is far more tameable than before, and on slow corners the car can be power oversteered quite predictably. As usual with Herald-based cars, the turn-on-a-sixpence turning circle converts most three-point manoeuvres into U-turns, and accounts for the high number of turns of the steering wheel (4.3) from lock to lock; the steering is actually quite direct.

As on earlier GT6s the brakes are a little heavier than is customary nowadays, making the pedal a firm pivot for heeling and toeing, but they gave a good maximum retardation, were progressive, and did not suffer from fade. The handbrake—no longer of the fly-off type—would barely hold the car on a 1-in-4 slope.

Comfort and controls

There is a Jekyll and Hyde character to the ride of the GT6. At high speeds on potholed roads with crumbling edges of the sort common in France the all-independent ride copes well with the irregularities and the car floats smoothly over the top. But at low speeds on the same surfaces there is some bounce and pitch and loud thumps when the wheels drop into holes; despite the radial ply Dunlop SPs there is also noticeable roar on smoother roads with fine-scale roughness. Mr. Hyde also appears on roads with transverse ridges such as the joints of concrete motorways, when the movement of the car becomes uncomfortably hard and bouncy and could, as a strong-stomached member of our staff remarked, lead to car sickness.

Putting the occupants' bottoms a little closer to the action by reducing seat thickness to provide much-needed extra headroom does not seem to have adversely affected comfort very much. The seats are reasonably comfortable, provide very good lateral support and have adequate range of fore-and-aft adjustment. If they lack thickness and softness anywhere it is on the backrests which are a little hard and need to give more lumbar support.

In most cases the advent of the safety regulations has brought about a marked improvement in the layout of facias and the new Mk. 2 dashboard is no exception to this, the controls of the much improved heating and ventilating system imparting a more balanced appearance to the array of instruments and knobs. The traditional Standard-Triumph side main-beam-through-dip stalk on the left has been replaced by a more conventional dip-flash-main beam stalk which is used in conjunction with a facia-mounted rocker switch; this is not as pleasant in appearance or action as those used by BMC. A right-hand stalk takes care of the indicators with the overdrive stalk right behind it which can make life a little confusing at times. Straightforward rectangular knobs—one with a pull and the other with a push and twist action—on each side of the wheel take care of the choke and wiping and washing. A further pair of rocker switches control the interior light and the electrically heated rear window, which is now a standard fitting. Major controls such as the handbrake, gearlever and pedals are all well located, heeling and toeing being particularly easy.

The crudely controlled heater of the older car has been replaced by a much more adaptable system with proper distribution and temperature control levers in conjunction with two fresh air vents on the facia and extractor louvres on the rear quarters. The heater output is good and there is a large enough flow of air through the very small interior to keep the car cool in hot weather; the throughput can be increased without much change in the noise level by opening the rear quarter-lights.

Much more noise, incidentally, was sometimes generated by the front windows which did not always wind up securely against their seals: when the sealing worked the wind noise level was moderate. Occasionally the screen misted up with one person inside during some steamy wet weather, but the fog was soon cleared up by the blower. The lift-up backlight with its built-in heating element showed less tendency to mist over. Forward and rearward visibility is good and with its compact size and superbly small turning circle, parking manoeuvres present no problem. The headlamps provided exceptionally good dipped and main beam illumination.

1, radiator filler cap. 2, air cleaner. 3, coil. 4, dipstick. 5, distributor. 6, oil filler cap. 7, clutch master cylinder. 8, brake master cylinder.

Fittings and furniture

Nicely trimmed and finished, the GT6 has a full set of instruments with rev-counter flanked by a matching speedometer containing total and trip distance recorders and fuel and temperature gauges. Hazard warning flashers, operated by a rocker switch, are a standard fitting. Above these instruments and controls is an ashtray—but there is no cigarette lighter to go with it—and beneath, an open glove compartment on the passenger's side with a vestigial one for the driver. There is more room for oddments and to conceal valuables in the compartments under the rear deck. Though there were no occasional rear seats at the back on our test car, small children are quite happy to lie or squat on this deck for short distances, or to sit on it with their legs dangling behind the seats when these are pushed forward a little. Luggage capacity is surprisingly good and the lift-up rear door an asset.

Servicing and accessibility

Even under a cramped bonnet the Triumph six-cylinder engine would be easy to look after, for most of its service points are readily got at. The distributor, for example, is located high up and on the opposite side of the engine to the carburetters rather than nestling underneath them as it does on one high-performance engine. But with the superb lift-up front, not just the engine but all the forward mechanicals are exposed so that it should be easy to carry out such work as changing a front suspension unit.

Servicing is required every 6,000 miles at which the main job is a change of engine oil; a little chassis greasing is needed at 12,000-mile intervals. M

MAKE: Triumph. MODEL: GT6 Mk.2. MAKERS: Standard-Triumph International Ltd., Coventry, England.

The Greatest GT6 in the World

DOWN AT the bottom of the garden behind a modern bungalow in car-crazy Coventry lies a perfectly ordinary double garage, which being not much more than a first-gear burst away from the Triumph factory, in normal circumstances might well be expected to hold a couple of that company's family saloons.

But if you'd opened the roll-over doors last year there'd have been quite a shock in store, for there, all alone except for a row of cupboards and a collection of International Rally plates, lay a sight to gladden the eyes of any enthusiast. An absolutely unrepeatable sight, in fact, for there lay all 333 bhp per ton of the only lightweight Racing Triumph GT6 in captivity — one of the most exciting and beautifully-prepared modified production sports cars to brighten the circuits for many a long day.

Not so much a phenomenon as you might expect, for the garage's owner happens to be a certain Peter Cox of Racing Spitfire fame, and the happenings at the bottom of his garden are usually far from normal, as the number of shattered eardrums amongst his neighbours will testify.

Over the last five years Peter Cox has made his name as the only person to make Spitfires competitive on the circuits outside the USA, first of all with his own ex-Val Pirie car with which he won the Freddie Dixon Trophy in 1967 and later preparing the extremely competitive Gold Seal-sponsored Spits for Richard Lloyd and Chris Marshall. The GT6, which he drove himself last season, was a logical progression.

As he spends his working hours on future projects and development for Triumph and used to work in the old Triumph Competitions Department in the days of those fabulous Rallying and Le Mans Spitfires, Peter is acknowledged as the leading expert on Spitfire preparation, though his Racing is completely private, with no works aid. So it's not surprising that the GT6 is something rather special.

And special it is — an 11 cwt, 183 bhp projectile capable of burning the freshly-laundered pants off even a full-house works 911S, lapping the Silverstone GP circuit in two seconds under the lap record on its first sight of tarmac while running-in during private practice, and in its first Race, at Thruxton in April, cheekily starting from pole position after beating Cobras, E-types and other hairy beasts in practice. Unfortunate, though, that the GT6's tender backside took a liking to the Armco at the Chicane when Peter missed a gear, robbing him of a possible first-time-out win.

Shortage of funds, ill-luck and, as Peter readily admits, his own lack of practice after a long lay-off from Racing, meant that the season didn't work out quite as successfully as might have been hoped. The intention had been to do a fair bit of Racing on the Continent, but in the end the car saw only six events after Thruxton, all on home ground. Not that one first overall, two firsts in class and two seconds in class can be called unsuccessful, particularly his showing in the combined Modsports/GT Race at the Silverstone AP meeting when the GT6 was the first Modsports car home and won its class. Last Race was the Thruxton 100 mile event when a blown head-gasket forced retirement.

Worst tragedy of the season was Mugello. Peter had asked Colin Malkin to share the drive with him, so off they trailed to Italy with the GT6 behind Peter's Jag 3.4 barge. Between them they knocked up some 80 miles of practice with the GT6 performing perfectly. Then the night before the Race Colin wrote off the Jag against a Fiat full of Italians, and both he and Peter were pretty badly cut about. Sickened though they were, they rolled up at the start of the Race with the GT6, only to be refused permission to start because of their injuries.

The saga of the GT6 started way back at the end of the '68 season when Cox did some serious thinking about the future. Seeing all those GT6s around at work every day with beautifully streamlined tiny bodies and comparatively big engine lumps, he wondered why the hell he was playing around with basically the same body yet a pint-sized engine, whereas a fully sorted GT6 could have a fantastic power-to-weight ratio. Before many pints had passed, there, sitting in his garage, was a GT6 chassis and the tedious business of acquiring, modifying and making the necessaries to bolt on to the chassis had begun. He and his mate from Triumph Experimental, Peter Clarke, who'd helped him through the Spitfire Racing years, set to with the midnight oil to complete the car for the '69 season, but found that the Spitfires they were preparing at the same time were taking too much of the oil. Rather than do a bodging job on all three cars they decided to concentrate on the Spitfires and shelve the GT6 till the following winter.

So as Spring burst forth last year, there bloomed in the Cox garage what must be one of the most potent and certainly the most beautiful Group 6 2-litre prototype on the Racing scene, hand built to the last rivet by the man who knows more about modern Racing Triumphs than anyone else who wields a spanner.

How on earth does this Cox machine turn the scales at only 11 cwt. when a standard car weighs over 17 cwt.? For a start it's all aluminium or glassfibre, only the chassis and screen surround having any relationship with the original tinware. Panels are the only remaining works Spitfire alloy ones: wings, doors, bonnet, bulkhead and floor pan, bought from the works along with a glassfibre roof section (ex-Le Mans Spitfire type).

No sticking this lot together with the old gas bottles — panels had to be hand-riveted together, such a tedious job Peter still has nightmares about it. Then there had to be a hacking job on the bonnet to slot it over the long six-cylinder mill, covered over with an aluminium power bulge hand beaten to GT6 shape and riveted into place, along with a smaller bulge to take the front Weber. Glassfibre spats, wide enough to enclose the 8½ in offset Minilites suitably clothed in Dunlop CR 82's, curve round and down at the pointed end to act as spoilers. First time out at Thruxton a Perspex panel filled the gap where the GT6 tailgate normally fits, but now a flat, glassfibre deck and sloping rear window, Lotus Europa-like, fills the space, though the side contour's still pukka GT6. Two huge Monza-type alloy filler caps mounted on the deck reveal quick-fill holes to the twin 10-gallon Autovita rubber fuel tanks separated from the tiny cockpit by a solid, fire-proof bulkhead. A single Bendix fuel pump between the tanks feeds the thirsty Webers.

Inside that Cox-sized cockpit everything's starkly, neatly functional, trimmed with the ubiquitous matt black paint. Driver's perch is a comfortable Restall (ex-works Spitfire again), with just a glassfibre pan to satisfy the regulations for the passenger who'll-never-be. If Peter is ever tempted to skate down the tarmac on the glassfibre lid, he'll always have the presence of a stout roll bar to comfort him, welded to the steel 'B' posts with an extension forward to the screen rail. Added to which a full Britax harness keeps his nether regions well and truly in the seat. A spare Minilite can just be squeezed in and secured behind the seats to suit International regs.

Again purely functional, the alloy facia carries only essential tell-tale dials: rev counter, oil pressure and water temperature gauges. A large area brake pedal (and it needs a heavy boot), fire-extinguisher, windscreen washer bottle (always thought those were for rally cars!), 13 inch leather-rimmed steering wheel by Bosmac, a Coventry firm, perspex side and rear windows with permanently open rear quarter-lights to reduce the claustrophobia, complete the picture.

Open that shapely bonnet, which incorporates the normally separate front valance and allows the removal of the valance-carrying

Sorting out the quickest GT6 of all at a Silverstone 'runing-in' session. Man in overalls is the Cox-man himself and t'other is Peter Clarke. Blonde bird is Mrs Cox. Job on hand in the bonnet-up photos is jetting up the Webers — a matter of trial, experience and a board full of spare jets. Carbs sorted, it pulled 7000 in top down the straight — 127.7 mph!

chassis extensions, and the most important operation in modifying this way-out car misses you completely unless you've got the clear mind of a strict teetotaller. That heavy six-pot lump is nine inches further back in the chassis than Triumph ever designed it to be! And with that, away goes the biggest handling detraction of the standard car — with a 55:45 weight distribution your common-or-garden GT6 carries a bit too much iron over the steered wheels, but the Cox blacksmithing gives an almost ideal 50:50 distribution. Yet on the face of it this seems to have been a fairly simple mod to perform: normal mounting points are to bosses on each side of the block, but by using a Spitfire front bearer plate on the front of the block, mounting on to the normal chassis points (with an extra inch welded forward), that magic nine inch movement is completed. Plus moving the rear gearbox mounting member nine inches further back in the chassis, chopping a piece out of the gear linkage and shortening the prop-shaft.

The Mark 2 rear suspension improvements on the standard GT6 tool transformed what had been quite a vicious, if not lethal, little beast in Mark 1 form into quite a desirable and predictably handling sports car. Needless to say, Peter Cox has improved on it even further — we say improved on it, when in fact he designed and made the same basic setup used on his GT6 for his team Spitfires in early '68, months before Mark 2 production cars appeared. The production banana-shaped bottom wishbones with single inner fulcrum points on the chassis are replaced by double stainless steel wishbones, cross-braced for rigidity, with two inner fulcrums for each wishbone pivoting on special nylon-type bushes. Radius arms are abandoned — all very well mounting on to a steel body, but mount them on to this shell and you'd end up with large lumps of ally coming adrift.

Suspension uprights are basically latest GT6 type, fashioned out of aluminium instead of malleable iron. Tempered Springs of Sheffield made up a flat transverse spring with six leaves instead of eight. Solid bushes replace the Metalistic spring end pivot bushes. Dampers all round are Armstrong adjustables, with that handy side ratchet tweaking device. Rear wheels have two degrees of negative.

With all the power that has to be transmitted to the rear wheels, driveshafts had to receive very particular attention — and treatment. Gone are those dreadful rubber doughnuts and in their place are Hardy-Spicer couplings grafted on to the outer ends of the shafts, almost countersunk into the uprights, while the rest of the movement is taken care of by sliding splines linking the inner and outer shafts. Diff is a 3.63:1 Salisbury limited-slip in the standard housing.

Front suspension mods are amazingly slight: wishbones, bushes and uprights are standard; a fat one-inch instead of three-quarter-inch roll bar, 7 inch instead of $8\frac{1}{4}$ inch long springs with thicker coils (again made by Tempered Springs and as used on the Racing Spits) with an alloy spacer to adjust the ride height, and the Armstrong dampers, are the only deviations from standard. Lack of adjustment facilities dictate the $1\frac{1}{2}$ degrees negative camber and a future project might be to make the front end fully adjustable.

Girling AR aluminium calipers force Ferodo DS 11 pads on to standard size (9.7 inch) discs made with a different offset so the bigger calipers will clear the Minilites. Rear drums are standard 8 x $1\frac{1}{4}$ with VG 95 linings. Girling tandem cylinders provide a dual system and there's no servo. So far the brakes have proved adequate, even over the punishing Mugello circuit.

Now we know how this exquisite machine contrives to corner so well, weigh so little, go so ridiculously quickly yet manage to stop, but where the hell does all that power come from? And 183 bhp at the flywheel at 7000 rpm is power with a capital 'P' when it comes from a 2-litre Triumph engine. To put it in proportion, it means some 20 bhp advantage over the 2.5 litre Weberised engine in Culcheth's Group 6 Scottish Rally-winning saloon and nearly 60 bhp more than the detuned 2.5 injected saloons which did so well on the World Cup Rally! Yet there could be still more power to come: triple 42 DCOE twin-choke Webers feed fuel to the six thirsty pots at present, giving more power lowdown than would 45s, but limiting top end power

40▶

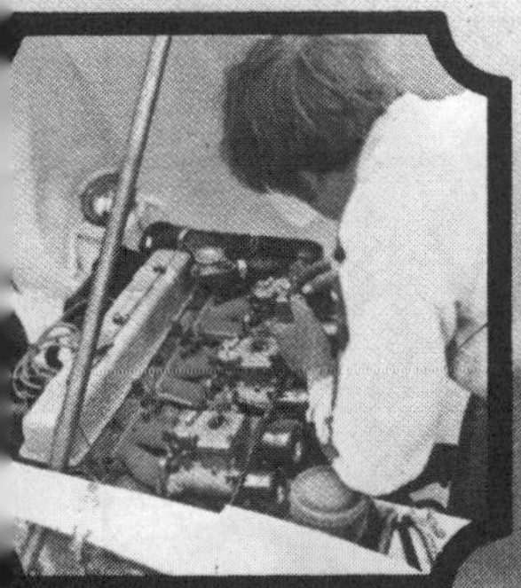

PHOTOS: ROBIN REW

39-10-4

Better roadholding for a popular sports coupé

THE Triumph GT6 is a very small car, closely related to the Spitfire. Instead of the 1300 cc engine of that car, however, it has a six-cylinder unit of 2 litres capacity, which has been cleverly shoehorned into the very short chassis of only 6 ft 11 ins wheelbase. The waisted chassis frame of Herald ancestry is retained, with a wishbone front suspension system and rack and pinion steering giving the famous minute turning circle.

It is at the rear that the big change has come about. The transverse leaf spring still spans the suspension assembly, but it now forms the upper members of the linkage, the half shafts being no longer employed as swing-axles by virtue of extra universal joints. The outer universals are of the rubber doughnut type, which are able to accommodate the plunging movement of the shafts. Lower reversed wishbones have been added to take over the duty of lateral location, and they are bowed to miss the doughnuts. The existing radius arms locate the wheel hubs fore and aft.

The well-known six-cylinder engine has been developed to produce an extra 9 bhp and, whereas the GT6 would never quite get up to 110 mph, it now does so with perhaps a little to spare. The Mk 2 version has an improved interior, the instrument panel being much more attractive, and better heating and ventilation are incorporated.

A big engine in a small car has always been the passport to effortless motoring. In the case of the GT6, the car is really tiny—indeed, it made the Fiat 850 Coupé in my garage look quite large. Two big people completely fill the living space and there is not much elbow room, while some contortion is called for when entering and leaving the carriage. Nevertheless, the luggage space is remarkably roomy, the rear window opening upwards for access, through which my two large dogs leapt nimbly—a recumbent human could travel thus for local trips.

Sharing most of its body panels with the Spitfire, the GT6 is very compact for a 2-litre car.

39-10-1

The GT6 is very much a sports car when judged by its appearance and driving position. It can also be regarded as a very flexible touring car, picking up from almost a walking pace to over 90 mph on the direct top gear. The overdrive is normally used a good deal, for the engine sounds somewhat busy at the higher speeds. This is something of an illusion, however, caused by packing a big engine and two people into a very tiny parcel. For this reason, the Triumph is noisier inside than a more spacious saloon but the sound level is not excessive for a sports car.

Wheelspin is easily induced and the car is lively, as would be expected. The gearbox is quieter than earlier examples and the changes go through very quickly. The gear-lever is somewhat stiff and sticky in action, but friction may have been introduced deliberately to stop the lever from rattling, a common fault from which the GT6 does not suffer. Though overdrive third and direct top have exactly the same maximum speed of 98 mph, the use of third and overdrive third on winding roads justifies this gear duplication.

On the road, the small size makes the car ideal for the rapid negotiation of traffic. With its low roof, it feels a bit beetle-browed at first, but one soon gets used to this. The sense of power in reserve makes this a very effortless car for long journeys, most other vehicles being smartly left behind with a touch of the accelerator. On direct top gear, there is a feeling of life and responsiveness which is delightful, the car remaining surprisingly flexible even when the overdrive is switched on at low speeds. Incidentally, the overdrive stalk is behind that for the flashing direction indicator, so it would be better if they were of different lengths. When crawling in bottom gear, the car tends to surge somewhat, due to the wind-

ROAD TEST

by John Bolster

Triumph GT6

With twin Stromberg carburetters, the 2-litre six produces 104 bhp.

39-10-2

Instrumentation lacks oil gauge and ammeter, and elbow room is limited.

39-10-3

SPECIFICATION AND PERFORMANCE DATA

Car tested: Triumph GT6 Mk 2 sports coupé; price, including overdrive and PT, £1211 17s 9d.
Engine: Six cylinders, 74.7 mm x 76 mm (1998 cc). Pushrod-operated overhead valves. Compression ratio, 9.25:1. 104 bhp (net) at 5300 rpm. Twin Stromberg carburetters.
Transmission: Single dry plate diaphragm spring clutch. 4-speed all synchromesh gearbox with short central gearlever, ratios 1.00, 1.25, 1.78, and 2.65 to 1. Laycock-de Normanville overdrive (extra) ratio 0.80 to 1. Hypoid final drive, ratio 3.89 to 1.
Chassis: Separate steel chassis frame. Independent front suspension by wishbones and helical springs with anti-roll bar. Rack and pinion steering. Independent rear suspension by transverse leaf spring, lower wishbones, and radius arms. Disc front and drum rear brakes. Bolt-on steel wheels, fitted 155 SR13 Dunlop SP68 tyres.
Equipment: 12-volt lighting and starting with alternator. Speedometer. Rev-counter. Water temperature and fuel gauges. Heating, demisting, and ventilation system. Flashing direction indicators with hazard warning. Windscreen wipers and washers. Radio (extra).
Dimensions: Wheelbase, 6 ft 11 ins; track (front), 4 ft. 1.25 ins; (rear), 4 ft 0.25 ins; overall length, 12 ft 3 ins; width, 4 ft 9.25 ins; weight, 17 cwt 3 qtrs.
Performance: Maximum speed, 110 mph. Speeds in gears: Direct top or overdrive third, 98 mph; third, 78 mph; second, 55 mph; first, 38 mph. Standing quarter-mile, 17.2 s. Acceleration: 0-30 mph, 3.4 s; 0-50 mph, 7.2 s; 0-60 mph, 9.5 s; 0-80 mph, 16.8 s; 0-100 mph, 29.8 s.
Fuel consumption: 25 to 28 mpg.

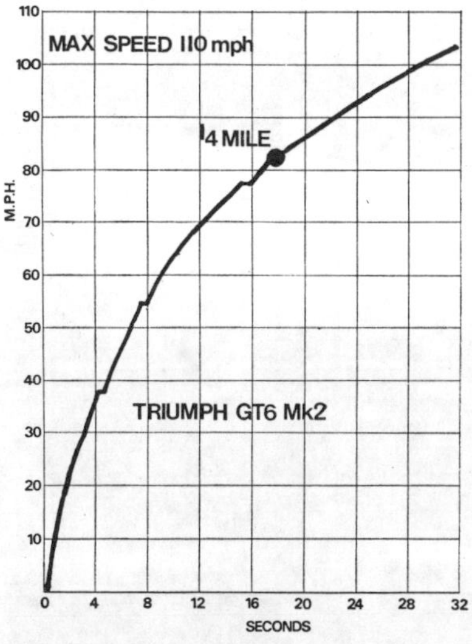

ing up and unwinding of the rubber doughnuts, a well-known characteristic of these joints on other makes of car.

The exhaust silencer looks as though it had been hung across the back of the car as an afterthought, but it does its job well. There is never too much noise but the glorious exhaust note is the pure music that only six-cylinders can give, and you can keep your fours and eights.

The GT6 is normally quite a strong understeerer but the tail can be broken away under power on the slower corners. It is remarkable how steady the short little car is at speed and, though the cornering is not quite in the racing category, the machine gets round most satisfactorily with very little roll. The steering feels rather dead and is not as light as would be expected, but it gives a good sense of control when the car is driven hard.

Personally, I would like to add a servo to the brakes, which lack bite for emergency stops and call for rather high pedal pressure. They are capable of pulling the car up quite surprisingly quickly, though, if the pedal is pressed hard, and they do not suffer from fading during fast driving. The parking brake is effective if the rigid central lever is pulled on smartly.

The suspension is fairly hard but free from pitching. On some types of road surfaces vibration can be felt both through the seat and the steering wheel, but road noise is not excessive and there is remarkably little wind noise. The seats give very good lateral location and though the padding is thin they are quite comfortable. Effective heating and ventilation are provided though some unwanted heat comes from the transmission tunnel in very hot weather.

The Triumph GT6 is a sports car with a roof. It is pretty and small, with lots of performance, and is a most attractive toy as well as being a practical, long-distance touring car. It is economical at quite high cruising speeds and is great fun to drive. The new rear suspension has made the Mk 2 version a much safer car for the less experienced driver to handle, and if some of the boys may miss the tricks they used to get up to with the swing-axle rear end, it is definitely a better car for all normal driving. Finally, the test car had an accurate speedometer, which is rare indeed.

GIAN
TES

GT6 v MGB

THOUGH IT WILL NOT ACTU- lie down, the whole concept he open sports car is pretty well d. The end will not come, unless legislation, so long as there are a long-suffering diehards around to and that they should be buffeted wind, irritated by draughts and alter- vely chilled or frozen—and always ened—as they drive along increasingly vded and polluted roads. For the rest of ess enthusiastic though we may appear, ast a modicum of comfort is desirable.

ence the growth in popularity, indeed virtual takeover, of the GT coupé in this . In northerly climes, at least, even the ting motorist rightly prefers a rigid over his head, and whereas at one time a er had to go to the likes of Jaguar, Ferrari or at any rate AC to combine these requirements, a sports coupé can, of course, now be bought from much more mundane suppliers who cater for the masses of ordinary people requiring lively, practical transport.

British Leyland, for example. The MGB has been the best-selling car of this marque ever. Its coupé version, the GT, has done equally well. Up the road at Triumph the little Spitfire-derived GT6 is also selling well and, like its MG cousin, it personifies the competitively priced, if in some ways rather dated, coupé two-seater. Both cars have more claim to those ubiquitous GT initials than do the saloons on which they are so freely used. As we shall see, they are fairly well suited to the Grand Tour—if not in luxury, then at any rate in comfort.

Only £87 separates these cars with the MG at £1356 and the Triumph at £1269.

The history of the MGB stretches way back into the dim and distant days of the early 1960s. The roadster was introduced in September 1962 and the GT variant followed in October 1965. Triumph's GT6 is similarly well established. It was introduced in 1966 in a rather primitive form.

Prices in both instances have unavoidably crept up over the years. Some increases, though, have been more justified than others. The MG last went up (by £57) in October 1970, while at the end of that month the GT6 was increased by £77. The rises represented soaring labour and material costs rather than worth while improvements for in both instances the 1971 alterations were no more than superficial.

Those to the GT6, in fact, were oddly timed. The Mark Three missed the Paris and London Motor Shows altogether and just

made Turin Show—the least important for Triumph. Even catching that was a struggle, it seemed, and two months later production examples were still as rare as steak tartare at a vegetarian picnic.

The last-minute decision to make a Mark Three also caught the marketing people flatfooted for their 1971 sales brochures refer to the GT6 Mark Two.

STYLE AND ENGINEERING

Alterations to the GT6 are minor enough to hardly make a mention, the main one being the adaption of the styling changes already planned for the closely related Spitfire. The chrome-trimmed raised seams along the tops of the front wings were deleted

Current batch of minor changes have done little to modernise the appearance of the MGB GT (top left) whereas the Triumph GT6 has benefited substantially from its latest modifications, including deletion of old-fashioned styling gimmicks like louvres. Cockpit of MG (centre left) has its comfort lessened by poor ventilation, although child seat in the back provides a scoring point over the strictly two-seat GT6 (centre right) which is generally more habitable. Long established B-series engine in MG sports a pair of SU carburettors in cluttered compartment (above left) whereas GT6's Stromberg-equipped engine is much more accessible

ın favour of a single panel with a razor edge along the top. On went a Stag-like tail; off came the old-fashioned louvres in the bonnet, the windscreen was deepened fractionally and the front wheel arches swaged out. Frontal styling was cleaned up and the imitation-Rostyle wheels were replaced by the ones from the TR6.

Mechanically the GT6 was unchanged, leaving it with its separate Herald/Vitesse-inspired backbone chassis, steel body, six-cylinder two-litre engine and all-independent suspension. Neither of these last two items is exactly the last word in design. The pushrod engine—a destroked, twin carburettor edition of the fuel injection 2.5litre unit in the TR6 and 2.5PI—is directly descended from that of the old Standard Vanguard, while the suspension utilises conventional double wishbones and coil springs at the front in conjunction with a lower wishbone/upper transverse leaf spring arrangement at the rear. This does give something approaching double wishbone geometry and is thus greatly superior to the horrible swing-axle system it at last superseded.

Sports car designers cannot fairly be criticised for having to make do with saloon-car components. The problem is the same at MG where the B has to use a preponderance of saloon bits and pieces, even though it does start off with its own body/chassis structure. In fact the B was the first unitary construction MG. Its four-cylinder engine is easily recognisable as the old BMC B-series, nowadays with five bearings to the crankshaft. It's still breathing through the traditional twin SU carburettors, prodding open its valves with pushrods and, like the GT6, does not exactly cover itself in glorious efficiency.

To be precise, the Triumph develops 98bhp from 1998cc, or 49bhp per litre. The MG produces 95bhp from 1798cc, or 53bhp per litre; neither are startling.

Direct comparison of this duo shows up a wide choice in terms of interior accommodation, although both turn out to utilise space with roughly similar effectiveness.

The GT6 is the smaller, showing its Herald/Spitfire origins in the stunted 7ft 11in wheelbase thus giving away nearly four inches to the MG. The Triumph is also the narrowest, measuring a slim 4ft 10.5in which is 1.5in under the MG. And there is, naturally, less room inside. The GT6 is a two-seater pure and simple—and a rather claustrophobic one at that. Deepening the screen has done nothing to alleviate the beetle-browed effect from within, heightened as it is by the restricted cockpit width. Behind the seats there is only a recess (a tiny seat to fill it is an optional extra) that gives room to recline the squabs. It provides a useful adjunct to the main luggage space though, which consists of a carpeted platform stretching back to the tail of the car. In theory one could load this up to the roof, wrecking rearward vision and risking violent assault from luggage launched forward during a panic stop. We feel that the reasonable limit is level with the base of the side windows, as in estate cars, and therefore take our luggage capacity measurements accordingly. This gives the GT6 a respectable 13cu ft which puts the MG way behind at only six cu ft. But in the MG some of what could have been luggage space has been given over to a back seat. It's extremely restricted for both head and leg room and is only suitable for young children. Even they will need to be long suffering as well, for the seat is thinly padded and has a very erect back. Its best function is simply as an extension of the luggage space and to this end its squab folds down to provide a shallow retaining lip to suitcases.

The MGB has a long, broad bonnet for styling reasons. As a result, there is some wasted space in the engine compartment whereas the Triumph's six cylinders demand more length, although the narrowness of the unit means that, again, space is wasted. The entire nose section of the GT6—bonnet and wings together—lifts up to give access to the engine and front running gear. The MG has a purely conventional bonnet and less access to the works.

COMFORT AND SAFETY

We liked the front seats in both the Triumph (specially now that their backrests recline) and the MG. They provided lateral location against cornering forces as well as adequate lumbar support.

The heating and ventilation system in the MG is as antediluvian as it was when we first cursed it in BMC cars many years ago. While the heating is adequate—barely—when it finally filters through, the ventilation is practically non-existent unless one opens the old-fashioned quarterlights. Even if the system admits little air to the cockpit it quite efficiently admits rain. In the test car water dripped relentlessly on to the legs of both occupants.

With the GT6 we have no major quibbles. Heating and ventilation alike cope with the job competently enough. A heated rear window is standard in the GT6 and a £19.58 extra in the MG.

The Triumph made all the usual sports car noises—engine, transmission, exhaust—in fairly subdued form but the MG had a noisy exhaust and valve gear which an owner could rectify easily enough even if the factory does not bother. Otherwise fairly quiet. Both were free enough of road-excited sound through the tyres, although there was wind noise originating around the windscreen pillars. The GT6 gained a great deal in smoothness over the MG from its two extra cylinders. The MG's ride is firm yet far from choppy; the Triumph's is rather less good due to stiffer springing.

The rear quarters hamper visibility in the GT6, although the forward view has been improved slightly by the deeper windscreen and the recontoured power bulge in the bonnet. The MG is not particularly good, either. The difficulty here is the unfashionably low seating position coupled

	MG	TRIUMPH GT6
PRICES	At £1356 (or £1441 as tested with overdrive and heated rear window), the most expensive MG on the market, now that the MGC has gone. The price of the standard model makes it £137 more than the open two-seater MGB	At £1269 it is lower priced than the MG by £87. The GT6 comes midway up the three-strong range of Triumph two-seaters, costing £394 more than the Spitfire, but £183 less than the 2.5litre TR6
ACCELERATION from standstill in seconds		
FUEL	25 mpg overall ★★★★★ 29mpg driven carefully 280–330 miles range 12 gallons capacity	24 mpg overall ★★★★★ 28mpg driven carefully 220–260 miles range 9.75 gallons capacity
SPEEDS IN GEARS		
HANDLING	Adequate but not brilliant cornering power on Dunlop SP68 radials. Rather heavy steering contributes to a general lack of agility by sports-car standards. Strong understeer, switching predictably to oversteer near the limit. Wet-road handling qualities relatively good	Understeers on Dunlop SP68 radials with the transition to final over-steer coming late but almost abruptly, at which point it is easily mastered with none-too-light steering. Especially rapid and accurate response to initial changes of direction. Wet roadholding satisfactory
LUGGAGE CAPACITY cubic feet		

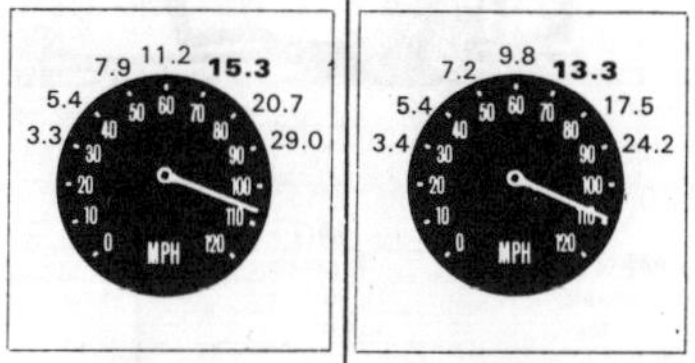

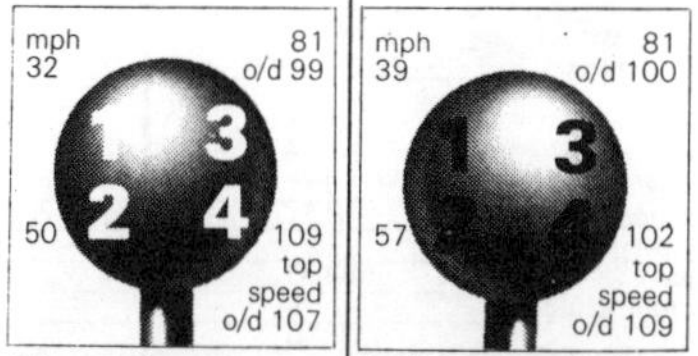

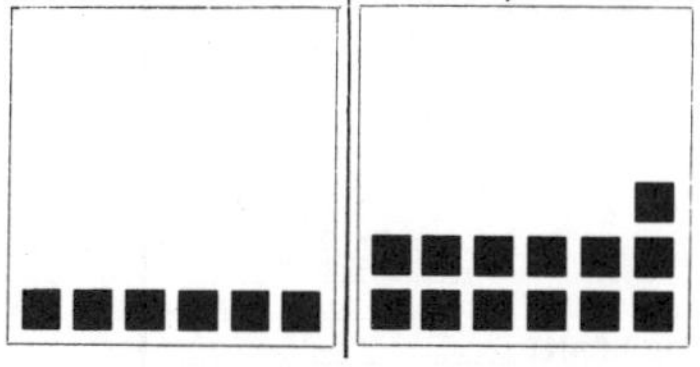

BRAKES RESPONSE in normal use. Deceleration (percent g) vs pedal load (lb)

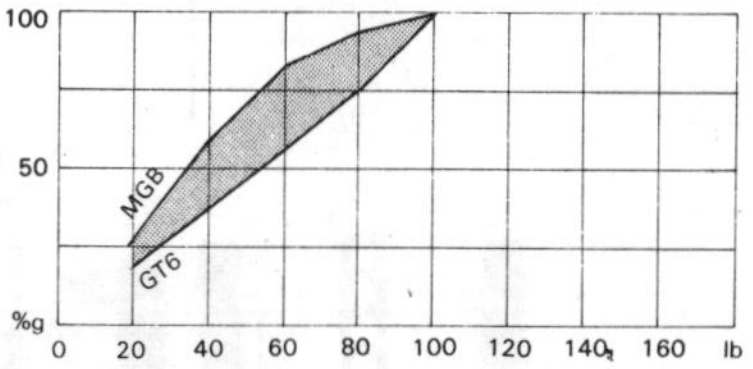

FADE peak deceleration achieved in 10 crash stops from 60mph at one minute intervals

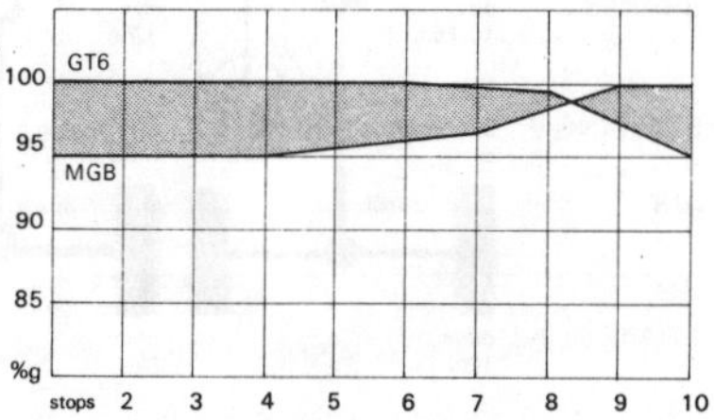

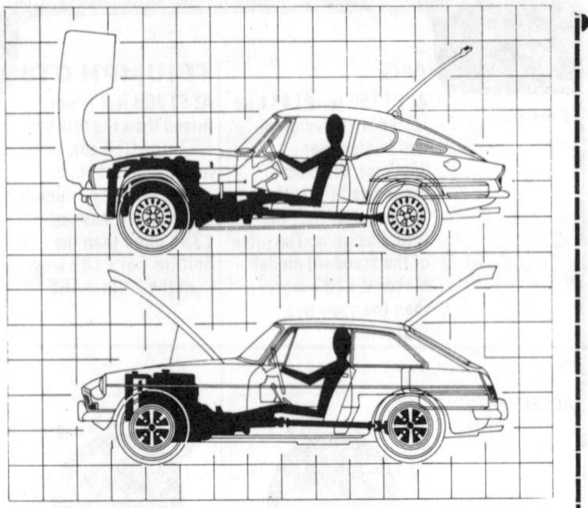

DIMENSIONS	MG inches	TRIUMPH inches
wheelbase	91	83
front track	49	49
rear track	49	49
length	152.5	149
width	60	58.5
height	49.4	47
ground clearance	5	5
front headroom	34	33
front legroom	27	30
rear headroom	25	—
rear legroom	3	—
ENGINE		
material	iron/iron	iron/iron
bearings	5	4
cooling	water	water
valve gear	pushrod ohv	pushrod ohv
carburettors	2 SU variable choke	2 Stromberg variable choke
capacity cc	1798	1998
bore mm	80.26	74.7
stroke mm	89	76
compression to 1	8.8	9.25
bhp	95	98
rpm	5400	5300
torque lb ft	110	108.4
rpm	3000	3000
TRANSMISSION		
control	floor lever	floor lever
synchromesh	1-2-3-4	1-2-3-4
ratios to 1 1st	3.440	2.65
2nd	2.167	1.78
3rd	1.382	1.25
o/d 3rd	1.333	1.193
4th	1.000	1.00
o/d 4th	0.820	0.80
final drive ratio	3.909	3.27
final drive ratio (with overdrive)	3.909	3.27
tyre size	165 X 14	155 X 13
rim size	5	4.5
SUSPENSION		
front	double wishbones, coil springs and lever dampers	double wishbones, coil springs and telescopic dampers
rear	live axle, semi-elliptic leaf springs and lever dampers	lower wishbones, upper transverse leaf spring and telescopic dampers
LUBRICANT		
engine oil type SAE	20W/50	20W/50
sump pints	8.25	8
change miles	6000	6000
other lube points	8	7
lube intervals	3000	6000

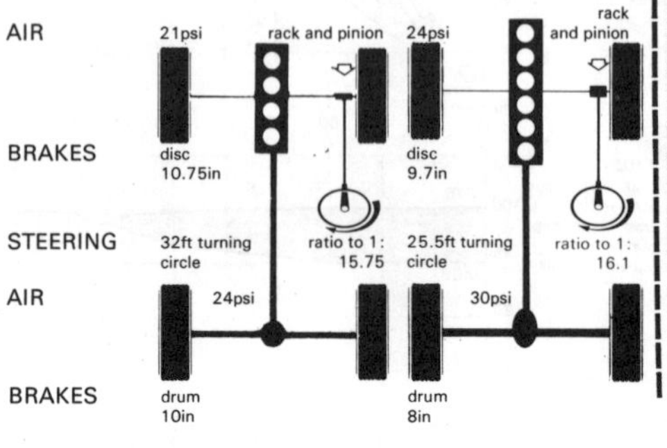

WEIGHT	2401lb	1972lb

with a high scuttle. For shortish drivers it becomes important that the windscreen wipers should park themselves flush with the screen surround. On the test car they did not and obscured the glass instead. Instrumentation was similar on both cars, comprising matched pairs of speedo and rev counter, plus other gauges for the essential services. The pair reflect the welcome trend away from banks of instruments, there only for their own sake, that were the hallmark of the sports car facia. Perhaps manufacturers could now cure themselves of their identical speedo and rev counter complex. It is all too easy, especially for a driver new to the car, to confuse them momentarily and since one tends to read an instrument by noting the needle position rather than actually scrutinising the figures this can be serious. The more thinking firms make the rev counter of a different size.

As with instrumentation, so we found too little to choose between the controls. Gear lever position and action was good on the GT6 and nicely placed but with slightly stiff movement in the MG. The GT6 steering column can be adjusted for length (by spanner work beneath the bonnet) and the pedals are considerably offset. In the MG it is impossible to heel and toe, which many sports car drivers will rightly consider a serious shortcoming that should have been rectified by the manufacturer; after all, the fault has existed for quite a few years now. Handbrakes are centrally located on both cars, the GT6 now having the lever and action used to such good effect on its stablemate, the Stag. Both the test cars were equipped with overdrive, controlled by a switch in the gear-change knob on the GT6 and by an absurdly placed facia switch on the B, but this tended to be offset by the annoyingly inaccessible ignition/steering lock—just about the worst we have ever encountered. The Triumph has recessed switches and controls on the facia but the Safety Fast MG, oddly enough, boasts a barbarous collection of rigid switches and other hard projections on the dashboard. We would expect the unitary construction of the MG to deform progressively under front or rear impact although the Triumph, with its separate girder chassis, would seem less likely to absorb collision forces in this way.

PERFORMANCE, HANDLING, BRAKES

The Triumph, gaining from its lower frontal area and lighter weight, starts to draw away from the MG once the 50mph mark is passed and is substantially quicker up to 90mph. Initial acceleration is not noticeably better in either car. The nose-heavy weight distribution of the GT6 counteracts the potential gain in adhesion from the independent rear suspension so that wheelspin is unavoidable during an all-out start. The Triumph should also have recorded a higher maximum speed, for it has the ability to do so on paper. In practice, though, it was no faster than the MG. So the pair put up respectable performances, but it should be remembered that neither are rapid in relation to their displacement. There are several saloons of similar engine size—albeit higher priced in some instances—which could keep up with them easily and a few could see them off. Fuel economy is quite good, being helped by the overdrive, although they require 100-octane petrol, making them more expensive to run than mpg figures suggest.

In their handling both came up to expectations being no better or worse than anticipated—satisfactory in other words.

The GT6's handling has been changed by the move from swing axles to primitive wishbone rear suspension. From awful to acceptable! It still understeers to start with, which exhibits much higher ultimate cornering power and eventually, almost abruptly, switches to oversteering. Directional stability has been greatly improved yet not to the point of woodenness: we liked the agile way the Triumph responded to an initial change of direction.

It did not share this characteristic with the MG. The overweight MG nowadays lags rather a long way behind in the handling stakes. Its cornering power is adequate but limited by rather pronounced understeer that makes it an extremely stable but hardly rewarding car to drive fast. At times the MG can feel almost cumbersome by any standards and certainly by those of sporting coupés. And added to this must be the fact that its steering is unnecessarily heavy. On braking, though, it put up as satisfactory a performance as the Triumph. Neither suffered undue fade and continued to pull up foursquare even under protracted duress, and pedal pressures were never unduly high.

IN CONCLUSION

Both the Triumph and the MG look the part of sports cars but have their work cut out to live up to it by present-day standards. The Triumph possesses a certain charm, due mainly to its compact build and handling qualities. It is quite well thought out and comes over as a distinctive car in its own right rather than the big-engined Spitfire that the appearance and specification might lead the cynical to expect. The recent changes, however, have done little more than improve the appearance without altering the basic shape.

The latest batch of alterations to the MGB GT have failed even to do that. In 1971 its looks, if not actually dated, are certainly established to put it kindly. And from the driver's point of view it is decidedly long in the tooth. Although the performance is not too bad for a 1.8litre car of this price and type, the handling is stodgy and marred by heavyish steering while the controls, facia and seating position are all reminiscent of the late 1950s. Worthy as countless owners consider it, the MG seems an uninspiring car today. ●

NEW CARS

Triumph Spitfire Mk 4 and GT6 Mk 3

Small Triumph sports cars revamped

The Spitfire Mk 4.

The Triumph Spitfire and GT6 have been restyled to follow the lines of the Stag. In addition, the suspension of the Spitfire has been modified to reduce rear wheel "tuck-in" and other swing-axle effects.

Starting with the Spitfire, it was decided that the more complex GT6 rear suspension, with double-jointed driveshafts, would make the smaller car too expensive. Accordingly, different steps were taken to cure the swing-axles of their bad habits. First of all the roll-stiffness of the rear suspension was reduced by mounting all the leaves of the transverse spring on a central pivot, except the main leaf, which is bolted to the differential housing. To compensate for this loss of roll-stiffness and to improve the balance of the car, an anti-roll torsion bar of greater diameter controls the double-wishbone post suspension.

The 1296 cc engine is basically the same, but larger big-end journals ensure longer life. A seven-blade fan and a new air-cleaner reduce noise, while quieter high-speed cruising is ensured by changing the final-drive ratio from 4.11 to 3.89 to 1. A new gearbox has a slightly higher bottom gear which is now synchronised.

The forward-hinged bonnet has a new shape, and the welded rib-joints on the front wings have been eliminated. There is a new matt-black radiator grille, with a chromium-plated wrap-around bumper mounted ahead of it. The rear panel is protected from rain and mud by a chromium surround, and there are new combined rear-light clusters. The forward view has been improved by a 2-in increase in the windscreen depth, and the optional hardtop now has opening rear quarter-lights. The interior has been completely restyled, with improved sound-damping, and there are new safety features.

The GT6, "the small car with the big engine," has only detail changes mechanically. Like the Spitfire, it has been restyled to follow the lines of the Stag, the whole front end having been cleaned up, and the wheel arches have been flared to allow the use of wider tyres if desired. By including the windscreen surround in the bulkhead assembly, the depth of the screen has been increased by 2 ins, and anti-lift wiper arms, as on the Stag, have been incorporated. The new rear-end treatment and wrap-around bumper are also reminiscent of the Stag.

The inside forward vision has been improved, as have the reclining seats. The shelf behind the seats has been removed to give greater luggage space and the starter control has been incorporated in a steering column lock. The capacity of the heating and ventilation system has been greatly increased, and there are new flush-fitting door handles with anti-burst locks.

I was able to test the Spitfire Mk 4 and the GT6 Mk 3 at Silverstone. The Spitfire is now very well-behaved, even when driven harder than would previously have been advisable. With very little roll, it fairly flies through the bends and handles far better than its predecessor. As a sports car it is greatly improved, but the new suspension gives a hard ride. This would not be detrimental on most of England's smooth roads, but there are still many parts of the Continent where the Spitfire would be somewhat uncomfortable. Comfort is seldom an attribute of small sports cars, but one of the advantages of independent rear suspension is usually the better ride it gives compared with a live rear axle.

The gearchange is very pleasant and the new box is quieter than the old one. The optional overdrive is operated by a very convenient switch on the gearlever knob and, though the car is greatly over-geared in overdrive top, it gives effortless high-speed cruising. The main value of the overdrive is to extend the range of third gear, to avoid changing into top on short straights between corners. The engine is fairly quiet for a sports car and runs easily up to 6,000 rpm.

The Triumph GT6 has a full 2-litre six-cylinder engine in a car the same size as the Spitfire, and naturally the acceleration is much more impressive. The engine has that delightful six-cylinder exhaust note and is remarkably flexible, putting its high gear ratios with ease. With its more complex rear suspension geometry, it handles just as well as the Spitfire but gives a more comfortable ride. The test car had a slight transmission vibration at about 95 mph.

The GT6 has a really vivid performance, with a maximum speed of 112 mph and a 96 mph third gear. The extra glass area of the screen makes the body feel lighter and more spacious inside. The small external dimensions are a great advantage in traffic and this is a car which is tremendous fun to drive, even on over-crowded roads. A large, torsionally-balanced rear door gives easy access to the luggage compartment. The brakes, formerly the Achilles heel of the GT6, have been improved both in power and stamina, and the car is of rugged construction throughout, engineered for continuous hard driving.

SPECIFICATIONS AND PERFORMANCE DATA

Car reviewed: Triumph Spitfire Mark 4 sports two-seater, price £962 including tax.

Engine: Four cylinders, 73.7 mm x 76 mm, 1296 cc. Pushrod-operated overhead valves. Compression ratio 9 to 1. 63 bhp (nett) at 6000 rpm. Twin SU horizontal carburetters.

Transmission: Single dry plate clutch. four-speed all-synchromesh gearbox with central remote control, ratios 1.0, 1.39, 2.16 and 3.50 to 1. Hypoid final drive, ratio 3.89 to 1. Extra: Laycock overdrive, ratio 0.802 to 1.

Chassis: Double backbone steel frame with outriggers. Independent front suspension by wishbones, coil springs and anti-roll bar. Rack and pinion steering. Independent rear suspension by wing and transverse leaf spring, all leaves except master centrally pivoted. Telescopic dampers all round. Disc front and drum rear brakes. Bolt-on disc wheels fitted 5.20S—13 tyres.

Equipment: 12-volt lighting and starting with alternator. Speedometer. Rev-counter. Water temperature and fuel gauges. Heating, demisting and ventilation system. two-speed windscreen wiper and washers. Flashing direction indicators.

Dimensions: Wheelbase, 6ft 1in; track, 4ft 1in; rear, 4ft; overall length, 12ft 5in; width, 4ft 10½in; weight, 15 cwt 1 qr.

Performance (makers' figures): Maximum speed 97 mph. Speeds in gears: third 78 mph; second 50 mph; first 31 mph. Standing quarter-mile 19.8 s. Acceleration: 0-50 mph, 9.0 s; 0-60 mph, 12.5 s. 0-80 mph, 23.0 s.

Car reviewed: Triumph GT6 Mark 3. Two-seater coupé, price £1269 including tax.

Engine: Six cylinders, 74.7 mm x 76 mm, 1998 cc. Pushrod-operated overhead valves. Compression ratio 9.25 to 1. 98 bhp at 5300 rpm. Twin Stromberg horizontal carburetters.

Transmission: Single dry plate clutch. Four-speed all-synchromesh gearbox with central remote control, ratios 1.0, 1.25, 1.78 and 2.65 to 1. Hypoid final drive, ratios 3.27 to 1. Extra: Laycock overdrive, ratio 0.801 to 1.

Chassis: Double backbone steel frame with outriggers. Independent front suspension by wishbones, coil springs and anti-roll bar. Rack and pinion steering. Independent rear suspension with lower wishbones, trailing radius rods and transverse leaf spring. Telescopic dampers all round. Disc front and drum rear brakes. Bolt-on disc wheels fitted 155SR—13 Dunlop SP 68 tyres.

Equipment: 12-volt lighting and starting with alternator. Speedometer. Rev-counter. Water temperature and fuel gauges. Heating, demisting and ventilation system. Two-speed windscreen wipers and washers. Flashing direction indicators with hazard warning. Reversing lights.

Dimensions: Wheelbase, 6ft 11in; track, 4ft 1in; overall length, 12ft 5in; width, 4ft 10½in; weight, 18 cwt 1 qr.

Performance (makers' figures): Maximum speed 112 mph; speed in gears: third 96 mph; second 68 mph; first 46 mph. Standing quarter-mile 18.2 s. Acceleration: 0-50 mph, 8s; 0-60 mph, 10.5 s; 0-80 mph, 18.5 s; 0-100 mph, 32 s.

The GT6 Mk 3.

ROAD IMPRESSIONS

GOOD VALUE FROM BRITISH LEYLAND-THE GT6

THE DAYS were not so long ago when the typical sports-car owner was that bachelor chap round the corner who was always rushing off to Silverstone with his bobble hat pulled down over his ears and a great roar of tyre smoke. The rest of the time he spent tinkering with the machine in some draughty garage getting incredibly dirty but achieving miracles with the engine. Now all that is becoming something of the past thanks to insurance companies and the like.

The present motor sporting fanatic probably does just as much rushing off to Silverstone and tinkering with engines but his car is no longer a sports model but more likely a Ford Escort with a big engine or a Mini with monstrously large wheels. The bobble hat has gone but there is still a uniform of sorts consisting of a "rally jacket" and a pair of kangeroo-skin gloves.

So who drives cars like the Triumph GT6 apart from all those Americans? The answer to this is simple. It is the chap who ten years ago would like to have been rushing around with a bobble hat pulled over his ears but, unfortunately, because he was either a student or an apprentice or about to get married he found it hard enough to finance the bobble hat let alone the sports car.

Now having made a position in the world and acquired on the way a wife, a young child and possibly a Siamese cat he hopes to recapture his youth. Naturally the sports car has had to change to meet the needs of somebody with such responsibilities and hence we come to British Leyland's popular sports car (call it a GT car if you wish)—the Triumph GT6.

MOTOR SPORT found the GT6 in its latest Mk. 3 trim a civilised, reliable, and also completely unfussy vehicle. We found the lines particularly pleasing and the performance well up to standard. We also realised that perhaps its biggest competitor is a fellow British Leyland stable mate the M.G.-B GT which sells for £1,356 compared with the GT6 which is a little cheaper at £1,287.

The car delivered to Standard House was a sparkling dark blue and turned out immaculately by the British Leyland Press fleet. Inside the interior was light brown and we immediately noticed the improvement in trim over the previous Mk. 2 model. When the GT6 was first announced back in 1966, we considered the styling to be a very clever adaptation of the Triumph Spitfire theme and despite a couple of face-lifts of a minor nature we still consider this to be so. Personally I do not like the nave plates on the wheels, which seem to be one of Lord Stokes' styling department's fads at the moment, but I suppose they are cheaper than hub-caps.

Our first task was to check the various levels and this is exceptionally easy with a GT6 as it is on the Spitfire and of course the Herald. The whole bonnet hinges up around the front, revealing all, most conveniently, and also reminding us that the GT6 still utilises a separate chassis. The straight-six 2-litre engine looks impressively powerful as it sits there. A similar engine still powers the Triumph 2000, although of course the TR6 and the 2.5 PI have now gone on to the enlarged engine with Lucas fuel injection. Everything comes easily to hand and if a mechanic wants to do any major task he can easily unbolt the whole bonnet assembly without too much problem.

On checking the oil we discovered that the car had been delivered

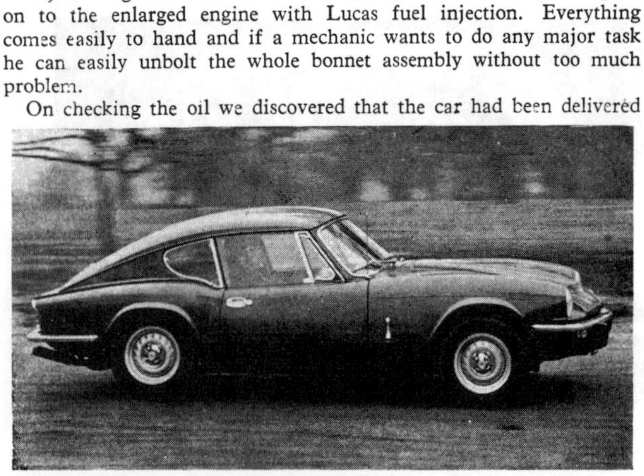

IN ACTION.—The Triumph GT6 looks effective at speed.

LIFT-UP TAIL.—This rear view shows the restyled rear end of the Mk. 3. Profiled in the rear window is the Lotus Elan +2S which will be featured in next month's issue.

with the level at the fill rather than full mark so we added a pint of GTX. At the end of our 900-mile test we needed a second pint to bring the car up to the full level again, so oil consumption will be around 900 m.p.p.

First impressions on driving a car are rarely, although occasionally, misleading and in the first ten miles or so one can usually get a fairly good idea of what a car is like. I usually reckon to make a quick mental calculation after a couple of miles which says "Do I feel completely happy and confident in this car to tackle the London rush hour now?". Whatever the answer I usually have to do just that but with the GT6 I felt at ease and comfortable too, but with a reservation or two.

Starting with the seating I found that there is plenty of adjustment both fore and aft and of the rake which any new owner would have to experiment. Personally I found that I needed a rather more upright position than normal as the steering wheel is rather large and being rather small in stature I found myself peering just over the top of it with an almost straight-arm driving position. There are covered sorbo-rubber pads on each side of the transmission tunnel so that both passenger and driver can rest their knees comfortably without knocking them. A central arm-rest, also well padded, makes for further comfort although it does not have any space to contain parking tickets and the like as the hand-brake sprouts out of the front of it.

Behind the two driving seats there is a spacious and well-carpeted area, but due to the way the rear sweeps down to the restyled tail there isn't the room to take an adult in any degree of comfort while small children would probably complain after a mile or two. However, that Siamese cat would have plenty of space, motoring dogs would also find the space acceptable, although MOTOR SPORT's example wasn't allowed to try this time, and there is plenty of space for baby's cot. The large rear window opens almost like an estate car, as illustrated in our photograph, and thus bulky parcels and so on can be loaded with ease.

The facia is wood (if it isn't genuine those plastic copies certainly look like it, these days) but the lay-out is not too bright. The centrally placed speedometer and rev.-counter are easily enough seen but what about the fuel gauge? This is badly obscured by the left hand in a ten-to-two driving position and one has to move one's head to view it. Although one can't see the situation of the fuel, how about checking that the oil pressure is nice and high? Bad luck again for the GT6 doesn't have such a gauge, simply an oil-warning light which, for a sports car, can hardly tell you if the pressure sags 10 p.s.i. when the car is driven flat out down the motorway.

While we are on the subject of little grouses how about the ignition switch and starter combination which isn't on the dashboard at all? That is buried down somewhere by your right knee alongside the steering column shielded by some padding in the shape of a cup. Finding the hole for the key is something like sinking a putt from the edge of the green. The reason the ignition is down here is obviously to do with steering-column locks, which I believe was a German idea and should really go the same way as Hitler. Quite honestly I think they are more trouble than they are worth. Still on this subject, if you ever break down in a car with such a lock and are about to be towed to your nearest garage remember to turn the ignition to unlock first; you would be surprised how many people have made that mistake.

In fact the keys of the GT6 confused us quite a bit. Both looked somewhat similar but the key that worked the ignition did not work the door locks. The key that operated them looked the same either way up but wasn't. In fact in one instance of fumbling to get the door open and the car started W.B. failed to escape the grasps of a parking warden. He swears that if it had been a one-key-for-everything Ford he would have been off and in 2nd gear before the yellow peril would have had time to raise a pencil to her notebook.

But these are only minor grouses; what of the general performance and handling of the car? The steering is tremendously light and responsive and something that could take a little acclimatisation particularly if one is used to something a little more solid. The leather-covered steering wheel was a little large for our liking and also rather hard to hold. Basically the steering is good and perhaps a slight drop in tyre pressure might make it feel a little more positive.

The gear-change, *via* a short stubby lever, falls easily to hand and was pleasant to use. One point here is that there is definitely a knack in selecting reverse, perhaps because our car was nearly new, but once learned there was little difficulty. Of the three drivers who tried the car all were unanimous about the brakes, they need high pedal pressure and lacked feel. At first when I drove the car I thought that the brakes were just plain poor. Then I started pushing harder and realised that it was all a matter of pressure. Rugger types would find no problem but the GT6 seems to find quite a good market amongst the female population of England and for them I think a servo would definitely be the order of the day. However, once you learned to push hard they did not fade at all and pulled the car up nice and square.

Unfortunately for the GT6 we had the vehicle at the same time as an Elan +2, of which much more next month, and our time was divided between the two cars. Lotus handling and cornering is legend and naturally, although the GT6 has fully independent suspension, it just doesn't come up to the standard of the much more expensive Elan. The old Spitfire transverse leaf rear suspension and tucking under rear wheels was thankfully not perpetuated on the GT6 and, of course, both the Spitfire and the GT6 now have a much more sophisticated method of springing the back wheels. The method works well and you can hurl the GT6 along at a good pace although naturally one doesn't have the same confidence as one would in the Elan. Round tight corners one can kick the back out quite easily although it is not a car I would reckon to throw about at faster speeds. To me it seemed a trifle unpredictable although this might be more due to a personal whim than pure fact.

The ride though a little choppy on rough roads is generally of a high order and cannot really be faulted, with the well-designed seats obviously helping here. Incidentally I particularly liked the seat belts which were well anchored and easily worn.

Although reasonably small and compact the GT6 with its steel body and separate chassis is no lightweight so the 2-litre engine does not give it sensational performance. Nevertheless it is a very brisk car which leaves the line very smartly and puts the power down well. The GT6 has a good top speed, too, although right at the top of the range one is starting to grit the old teeth a bit. Perhaps the best cruising speed is around 90 m.p.h., which the GT would perform happily all day and every day. At an indicated 100 m.p.h. the rev.-counter was showing 5,000 r.p.m. while it will continue to accelerate up to 5,500 r.p.m. which is where the yellow line starts, with the red around 5,800 r.p.m. In fact we were able to hold 5,500 which must be about 105 m.p.h. for quite a while although at that speed one is aware that the GT6 is close to its limit. However, at an indicated 100 m.p.h. everything was well in order and I was able to take my hands off the wheel without drama to note that the car ran straight as a die. Mid-range acceleration is also particularly good, for the engine has plenty of willing torque so one isn't forever having to change gear.

The heating and ventilating system proved to be efficient and easy to operate although the inclusion of a heated rear window would be almost essential. We were surprised to find that, actually under the facia, there were an additional pair of eye ball sockets to the ones on the main dash. These were so tucked away that we didn't find them until the test was half over, but the general idea seemed to be to blow warm air directly onto your cold feet.

As the miles mounted up I discovered a few more traits of the GT6 which had not struck me at first. One concerned the light dip switch which was operated by a stalk from the steering column. I don't reckon you can beat the present all-purpose stalks which are used these days by several manufacturers. Such a stalk gives you indicators left and right by up and down movement and lights full or dipped by forwards and back movements with a horn on the end tip for good measure. The GT6 had separate indicator and light stalks, both with up and down movements, but thankfully no overdrive or we really would have been confused.

The really broad-shouldered driver may also find that he is a little cramped by the GT6. The cockpit space is fairly small and I found my shoulder only an inch or two from the door.

In summing up the GT6 one must emphasise that British Leyland have produced a strong, reliable and well-engineered sporting vehicle which is, above all, very competitively priced. Neither Chrysler, Ford or Vauxhall offer a competitor and comparable sporting 2-litre two-seaters from the smaller or foreign manufacturers sell at a good deal more than £1,300. Running costs are not high either with fuel consumption working out at around 25 m.p.g. driven hard, and maintenance bills fairly reasonable too. The GT6 in its latest Mk. 3 form is known to be a well-sorted and bug-free motor car, and is tailored to meet the needs of the kind of person who buys it. We would not recommend it to the rorty-torty brigade who want to go hurtling about the country in opposite-lock slides with the engine revving round to 7,000 r.p.m.

However for someone who still likes the idea of a two-seater with excellent yet unfussy performance complete with all the mod cons of a well-finished interior, wind-down windows and so on there isn't much to beat the GT6. Undoubtedly one of British Leyland's better cars. — A. R. M.

THE BONNET hinges back to reveal the 2-litre six-cylinder engine.

COMFORTABLE SEATING and a rather large steering wheel are features of the GT6's interior.

Comparing the Datsun 240Z, Fiat 124 Sports, Opel GT, MGB GT and Triumph GT6—a closer contest than we expected

WHAT DOES ONE get when he buys $3500 worth of Grand Touring car? Generally, a small, light closed car, adequately but not spectacularly powered, a cut above the average in handling and braking, offering a measure of comfort not found in an open sports car. This month we take a comparative look at five such cars. None of them is a new model, all having been tested before by R&T in separate road tests. There have been detail changes in all of them since we last tested them although their basic character has not changed; still there's nothing like getting a group of cars together, taking a journey in them and comparing them nose-to-nose. We're always surprised at how much we learn in a comparison test and the reader may be surprised at some of the results of this one. We were.

As the General Data table shows, the list prices of the five cars all cluster around $3500. And each car is reasonably complete at the basic price—you don't need to pay extra for adjustable seats, a cigarette lighter or a 4-speed gearbox. All are front-engine, rear-drive cars, all have inline piston engines, all have at least two disc brakes out of four, and all weigh between 2000 and 2500 lb; one is from Japan, one from Italy, one from Germany and two from England. And they are all very different from each other.

The Datsun 240Z set U.S. motoring on its ear when it appeared in early 1970. It seemed too good to be true: a really fast, good-handling and good-looking coupe with great refinement and extensive standard equipment—all at what seemed an incredibly low price. Strictly a 2-seater but a roomy one, it is the longest car of the group though not the widest, the heaviest by a small margin, and by leaps and bounds the most powerful with 150 bhp from its 2.4-liter overhead-cam 6-cyl engine. It has independent suspension all around (by struts and coils at both ends) and a combination of disc front brakes and drums rear. Its styling is professional and up-to-date if a bit "pop culture," and that the 240Z is an exciting package cannot be denied.

In fact, it is so exciting that it has generated a supply-and-demand situation virtually unprecedented in the U.S. The waiting list for delivery on one is as high as six months, even though it has been over a year since the model was put on sale. Datsun programmed a supply of 1600 per month for the U.S. and they're now getting over 2500, but they have found that the demand is for about 4000 per month. Interesting, for we asserted in the 240Z introduction story that a domestic carmaker probably could build an equivalent car for the same price ($3600) in quantities of 50,000 per year or so. Anyway, Datsun stopped production on the 1600 roadster to make way for more 240Zs but the supply is still far behind. In some areas, for instance, dealers are telling customers they can't get the cars without lots of optional equipment—wide alloy wheels, air conditioning, etc., and getting away with it. And the *Kelley Blue Book* retail value for a used 1970 240Z is over $4000!

So—though the list price of a 240Z may be only $3596—one may or may not be able to get one for that price. If you can find a dealer who will sell you one for that price you'll probably have to wait months to get it. We must give Datsun the credit for producing such a package at such a reasonable price but we have to caution the reader that Datsun's price may not be the dealer's price. The laws of supply and demand still work, list prices or no.

Standard equipment on the 240Z includes a signal-seeking AM radio (with electrically powered antenna) and a heated rear window. Our test car had no options except a set of wide wheels (14 x 5½, part no. 40300E4600, $13.50 each) that can be obtained from Datsun dealers but are not installed at the factory; for the price table we estimate a total charge of $74 for the wheels and installation.

The Fiat 124 Sport Coupe has been updated this year with a smooth new front-end look and a 1608-cc engine of longer stroke than the older 1438-cc unit. It's not any faster than before but the larger engine is stronger in the middle ranges (so that less gearshifting is required to maintain a given pace) and both smoother and quieter. The Fiat is almost as long as the Datsun and somewhat wider; it is the only car in this group to offer a real rear seat and it's so real a seat that the car can be compared to some small sedans. It has several engineering distinctions: the only twincam engine in the group (the two camshafts are driven by a single toothed belt), the only 5-speed gearbox and the only 4-wheel disc brakes. It's the only car in the group with a separate, lockable trunk where one can hide valuables; three of the others have "tailgates" and one a cockpit luggage area. And its luggage capacity is the largest in the group.

The 124 Coupe test car, working from a POE price of $3292 (the lowest in the group), had the nice Cromodora alloy wheels, which cost only $135 for the set, a rather poor AM/FM radio (one shouldn't expect much at $85) and add-on chrome side strips and luggage rack; the last two items aren't included in our price tabulation.

British Leyland's MGB GT has been around since 1966

THE $3500 GT

and is based on the MGB roadster introduced in 1962. It's a classic British sports car but with a nicely designed fixed roof. The only one of the group attempting to be a 2+2, it has a very small bench seat behind the two main ones; this is big enough for small children only, and it can be folded to extend the luggage area. Surprisingly, the B is nearly as heavy as the Datsun, and with only 92 bhp from its pushrod four it is the slowest of the lot. Its front suspension is conventional—unequal-length A-arms and coil springs—and rear suspension is by the simplest means possible, a live axle on leaf springs.

The MGB GT comes with radial tires as standard equipment; styled steel wheels are standard, wire wheels (either painted or chrome) optional. Our car had the optional overdrive ($165), the familiar Laycock-de Normanville hydraulic unit that shifts in or out at a flick of a stalk on the steering column to give a 0.802:1 reduction. Thus 4th gear becomes 3.14:1 overall instead of 3.91:1 and the MG becomes the longest-legged of the lot. The overdrive works on 3rd gear also.

Opel's GT is based on the Kadett chassis, so it actually isn't as up-to-date in that department as its bulkier cousin the 1900 Rallye. Its front suspension is odd, based on a transverse leaf spring and one set of lateral control arms, but at the rear everything is shipshape with a live axle located by trailing arms and Panhard rod and sprung by coils. Its all-steel unit body (many people assume it's plastic because it looks so much like a Corvette) is made in France to a very high standard of finish and for 1971 it is available only with the 1.9-liter engine rather than both the 1.1 and 1.9. The engine has taken a power cut, though, to satisfy the politicians; bringing the compression ratio down from 9.0 to 7.6 so it could run on 91-octane fuel has reduced power from 102 to 90 bhp. Performance has suffered (0-60 time is up from 10.8 to 11.9 sec) but the GT still holds its own in the class. Amazingly, the GT has had a price cut too—it's now nearly $200 cheaper than it was in 1969 with the 1.9 engine.

The Opel GT is strictly a 2-seater: luggage is carried on a flat floor behind the passengers and loaded in through the car doors. The spare tire lives behind a vinyl partition just aft of the luggage area. Thus it has the least convenient luggage accommodation, though not the least capacious. Other distinctions for the Opel: it's the lightest and most economical of fuel in the group and its brakes can stop it in the shortest distance. Our test car had but one extra, a good AM

PHOTOS BY GORDON CHITTENDEN

$3500 GT

Datsun's engine is the largest and its instruments and controls the best.

Fiat has the only dohc engine and excellent instrumentation.

Opel's engine has generous displacement and unusual high-cam design.

MG has time-proven pushrod engine, safety-modified instrument panel.

GT6's smooth engine is very accessible, its wood-panelled dash luxurious.

radio at approximately $75 (dealer installed); like the Datsun, it can be ordered with a 3-speed automatic transmission.

Before the 240Z appeared the Triumph GT6 was the only car in its class with a 6-cyl engine, but now its 95-bhp, 2-liter six is no longer a big attraction. It's the smallest car of the group and only insignificantly heavier than the Opel; these points relate to the fact that it's derived from Triumph's smallest, lightest sports car, the Spitfire. It shares basic chassis structure (a backbone frame with separate body), front suspension and body structure from the beltline down with the open-bodied, 4-cyl Spitfire though the Spitfire's swing-axle rear suspension is replaced in the GT6 by a more satisfactory unequal-arm linkage for the rear wheels. The GT6 has the tightest interior dimensions of the group but not the smallest luggage capacity, and its luggage area is loaded easily through a tailgate as in the Datsun and MG.

This year the GT6 has been freshly restyled on the outside, and if some staff members noted that the rear end is reminiscent of the old Sunbeam Harrington coupe, all agreed that the car is better looking than before. It has one feature that makes it uniquely maneuverable in a crowded city: a tiny turning circle made possible by front wheels that can be steered well beyond the limits of proper geometry. One can turn around in just 25 ft, 6.5 ft tighter than the next twistiest car in the group, the Datsun.

Standard equipment at the GT6's basic price of $3424 includes white-stripe radial tires and a heated, tinted rear window; it has a 4-speed gearbox and an overdrive like the MG's can be ordered as an option; brakes are a disc/drum combination. Our test car had only an AM/FM radio—again a rather poor and low-priced ($100) one. A final distinctive feature of the GT6: A rich wood-panel dashboard in the traditional British manner.

As IN PAST comparison tests, we chose a test route appropriate to the character of the cars. In this case it was a route we'd used two years ago for a group of more expensive GT cars. We left our office in Newport Beach, topped up all the cars at a nearby filling station, and drove south

GENERAL DATA: 5 MEDIUM GTS

	Datsun 240Z	Fiat 124 Sport Coupe	MG B GT	Opel GT	Triumph GT6 Mk3
Basic POE price*	$3596	$3292	$3620	$3306	$3424
Price as tested	$3745	$3562	$3823	$3409	$3674
Engine position/driven wheels	f/r	f/r	f/r	f/r	f/r
Chassis type	unit	unit	unit	unit	separate
Brake type	disc/drum	disc	disc/drum	disc/drum	disc/drum
Swept area, sq in./ton	233	227	227	222	209
Suspension, front	ind. coil	ind. coil	ind. coil	ind. leaf	ind. coil
rear	ind. coil	live coil	live leaf	live coil	ind. leaf
Standard tires	175-14 rad	165-13 rad	165-14 rad	165-13 bias	155-13 rad
Steering turns, lock-to-lock	3.5	2.7	2.9	3.0	4.5
Steering index	1.10	0.99	0.93	0.99	1.13
Fuel tank capacity, gal	15.9	11.8	12.0	13.2	11.7

*POE prices vary slightly for east, west and Gulf ports

As-tested prices include: for Datsun, 14 x 5½-in. wheels and installation; for Fiat, alloy wheels, AM/FM radio; for MG, overdrive; for Opel, AM radio; for Triumph, AM/FM radio. All as-tested prices include charge for preparation at dealer.

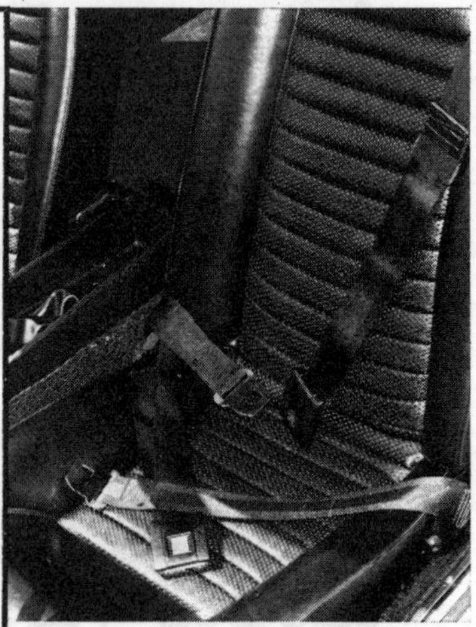

Opel's separate lap-shoulder belts work but are a mess.

on the Coast Highway through Corona del Mar and Laguna Beach (stoplights galore) to Dana Point, where we turned inland to connect with Ortega Highway, California 74. This highway, a nicely surfaced, 2-lane route with a delightful variety of turns and hills as it winds through some of Southern California's finest country, was lightly traveled and the weather was beautiful; it was easily the highlight of our trip. At Lake Elsinore we connected with route 71 toward Temecula; after Temecula, route S16 to Pala, 76 to Santa Ysabel and 78 to the little town of Julian high in the Laguna Mountains, where we stopped for lunch. Then on down 78, another wonderfully twisty road, into the Anza-Borrego desert, out across the desert at speeds dictated by road conditions and car capability rather than artificial limits, up through the Joshua Tree National Monument, and over clear, generally straight back roads to our overnight stop at Victorville, from where we freewayed it back to Newport Beach the next morning. In all, a rich and varied 500 miles of motoring in which we found out all about the five GTs.

Back at the office we set about scoring the cars. Each driver was given a score sheet on which he could rate each car on 15 different aspects of behavior, such as handling, ride, quietness, braking, steering, gearbox, engine, controls, seating, ventilation and heating, vision, finish, luggage accommodation and so forth. All these categories could be scored on a 1 to 10 basis, 10 being the score for a topnotch performance and 1 being the lowest score possible. These scorings were then totaled for each driver and for the entire group to get an overall rating score for each car.

In addition, each driver was asked to rank the cars in the order of his personal preference—disregarding, if need be, his separate ratings of the car's various aspects. Here's how the ratings turned out:

The Datsun scored the highest point total. In individual driver scoring, it garnered the highest number of points from four of the five drivers, and three of the five drivers rated it their personal favorite.

Next came the Fiat, and here we emerged somewhat surprised. It had been generally anticipated that the Datsun would win by a large margin, but not so. The Fiat tallied an impressive score, little less than the Z; one driver gave it more points than he gave the Datsun, and the same driver gave it his personal nod.

Then the Opel. It was a clear step below the Datsun and Fiat but clearly not in the doldrums. One driver rated it his personal favorite, though in scoring he had given the Datsun more points.

In total points the MG was not as far below the Opel as the Opel was below the leaders, and there was no unanimity in the personal ratings of the MG by the various drivers: two rated it third, two fourth and one last. But these ratings averaged a 4th-place finish just as clearly as the points score indicated; in fact, averaging the "position" of each car over the five drivers' listed orders of preference, the cars stacked up the same way: Datsun, Fiat, Opel, MG, Triumph. Which brings us to the Triumph: It came in last, but not far behind the MG and was ranked last by four of the drivers on their personal ratings. The one driver that ranked it 4th instead of 5th also gave it more performance points, so it had a clear attraction for him. Now let's look

GENERAL SPECIFICATIONS: 5 MEDIUM GTS

	Datsun 240Z	Fiat 124 Sport Coupe	MG B GT	Opel GT	Triumph GT6 Mk3
Curb weight, lb	2355	2220	2345	2110	2115
Test weight, lb	2770	2620	2725	2500	2490
Distribution, f/r, %	51/49	55/45	49/51	54/46	54/46
Wheelbase, in.	90.7	95.3	91.0	95.7	83.0
Track, f/r	53.3/53.0	53.0/51.8	49.3/49.3	49.4/50.6	49.0/49.0
Length	162.8	162.3	152.7	161.9	149.0
Width	64.1	65.8	59.9	62.2	58.5
Height	50.6	52.8	49.4	48.2	47.0
Luggage capacity, cu ft	8.5	9.6	6.3	6.6	6.6

MG's tailgate-loading luggage area is handy.

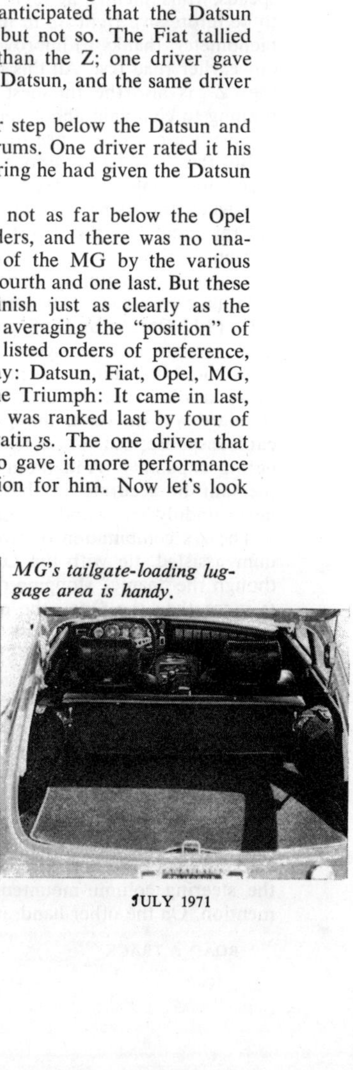

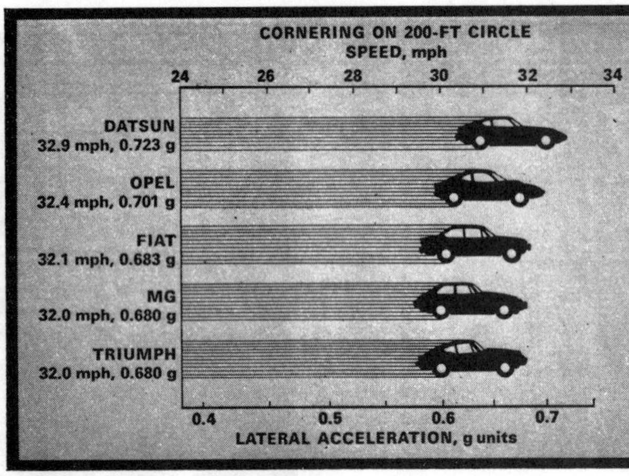

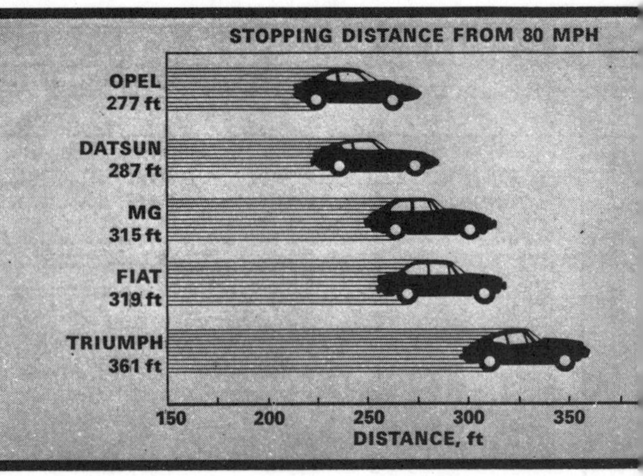

THE $3500 GT

at the cars in detail and in the order of their ranking.

Datsun 240Z

IN GENERAL the 240Z lives up to its promising specification. Its generous-size engine delivers the smoothest, quietest and dramatically the most powerful performance in the group; it is an engine that pulls strongly from low speeds, runs silently at high cruising speeds and continues to be impressive right up to its 6500-rpm yellow line on the tachometer, thanks to improvements in the crankshaft since our earlier road test. All this performance isn't without cost; the "Z" is also the thirstiest in the group, but 21 mpg is nothing to complain about and the fuel tank is large enough to give it a cruising range to match its cruising ability.

But the Z-car has a serious problem associated with high-speed motoring. It's very sensitive to sidewinds at speed, and when traversing road undulations at high speed (as we did repeatedly on the desert highways) it requires a lot of motion at the steering wheel to keep it on course. Datsun has some chassis tuning work to do here, and in the meantime an owner can fit an undernose spoiler, available from BRE (137 Oregon St., El Segundo, Calif. 90245) for $32.

In low- to medium-speed cornering and handling, however, the 240Z shines. Surprisingly, the 5½-in. rims didn't make a difference in absolute cornering power over the standard 4½-in. ones, though they added crispness to the car's response; but it still led the group with a 0.723g cornering capability. And though its steering is not the most pleasant or accurate in the group, it is quite acceptable, never unduly heavy and certainly quick enough.

The Z's combination of front discs and rear drums, vacuum assisted, tie with the Opel's brakes for best-in-group; though the "panic" stopping distance from 80 mph takes 10 ft more than the Opel, the car stays under control a little better under these conditions and fade under hard repeated use is negligible.

Comfort and accommodation also rank high in the 240Z. It accommodates only two people, but what space for those two! The largest driver in our test crew, who is 6 ft 2 in. tall and weighs 200 lb, gave it the full 10 points on every aspect of its interior but finish—undeniably some of the materials used, notably the quilted-pattern vinyl, are less than pleasing. Controls are notably good, everything being within good reach for a typical male driver with his 3-point belt fastened (the best belt in the group, by the way), and the steering column-mounted lighting control rates special mention. On the other hand, it was the only car in the group without some provision for easy daylight headlight flashing, and the ventilation system, though it provides a good flow of outside air that can be boosted by the blower, unfortunately aims most of it right between the driver and passenger.

In sum, the Datsun 240Z's plusses are its striking looks, its effortless, strong performance, its good brakes and low-speed handling, and its comfort and equipment. On the negative side, the only serious criticism is about the high-speed stability. If you can get one for list price, or even get one with the extras *you* want, it is not only the best car in the group but the best buy.

Fiat 124 Sports Coupe

THE FIAT deserves more popularity. At nearly $200 less than the Datsun with comparable equipment, it did so well in our comparison test that it scored nearly as many points. Of course it doesn't offer the zoomy styling of the Datsun (the boxy shape that turns it into a true 4-seater doesn't allow that) nor the brilliant performance. Its 4-cyl engine, the smallest of the group, is nevertheless a most satisfying bit of machinery: quiet, very smooth for a 4-cyl (easily the best four in the group), and willing to rev happily to its 6500-rpm redline. And the 5-speed gearbox is the best gearbox in the group.

In road behavior the Fiat scores at the head of the group. Its steering is the most precise, its handling the best; it really shines at high speed in contrast to the Datsun, for it isn't blown about by sidewinds and can negotiate high-speed dips and humps without a hint of losing its composure. Only an over-eager vacuum brake booster detracts from its overall roadworthiness; our drivers always found themselves overdoing it with the brakes when first getting into the car. And though it's the only car in the group with disc brakes all around, it doesn't do anything impressive in a panic stop and the brakes squeal often when not in use. Its brakes do have the best fade resistance in the group, though.

The Fiat's driving position is, in a word, odd—and perhaps something we, as Americans, will never understand. The steering wheel is buslike in that it is less vertical than usual, and it is far away from the driver while the foot pedals are close. But the seats are good and so are the controls, which make maximum use of modern steering-column stalks to do various things. The seatbelts were installed incorrectly on the test car so that either lap or shoulder section was twisted. Ventilation is not particularly good, but ventwings in the doors make it possible to drive at moderate speeds with the door windows open and no drafts. The Fiat has the best vision outward of any car in the group, so it's a good car for city traffic.

With its good rear seat and capacious, separate trunk the 124 Coupe is far and away the most practical car of the

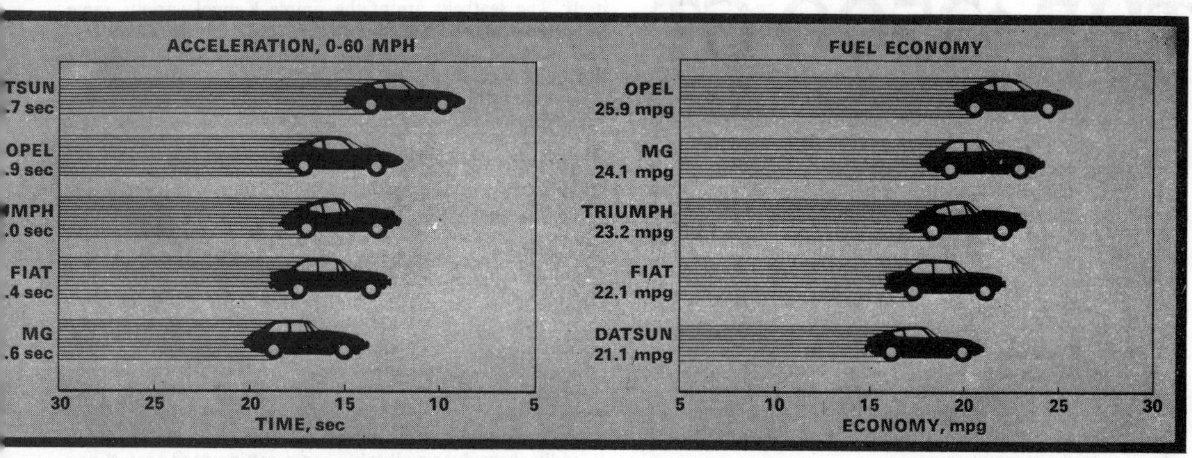

group if one needs more than 2-passenger accommodation. And surprisingly it doesn't give up a thing in sportiness for this extra measure of utility; in fact, it is *the* driver's car of the group. It is also well finished and equipped. On the minus side are its modest performance, grabby brakes and fuel economy that isn't impressive. We placed the Fiat 124 Sports Coupe as a close second in the group.

Opel GT

THIS ONE, coming in third in the group, is a crisp little package but not an impressive value for the money. Regardless of what one may say about its pseudo-Corvette styling, it is extremely well finished and put together and ranks fairly high in its driving position and control layout. It has the 2nd-best gearbox, light if not dead-accurate steering and the highest performance level of "the others" (i.e., other than the Datsun) despite its excellent fuel economy. It's the only one of the group that uses regular fuel.

The body is structurally solid and rattlefree. Vents at the dashboard ends provide a good flow of ventilating air, and longlegged gearing makes the engine (which is rather hashy sounding up through the gears) fairly quiet at speed in 4th; thus one can cruise with the windows up even in warm weather at an untiring noise level. On curvy roads the GT is amazingly well-planted and stable considering its humble origin, but in hard low-speed turns the rear axle chatters and its general handling characteristics are too much on the understeer side for maximum entertainment value. As we said in our road test of it, it may not be a really good-handling car but it is safe and predictable.

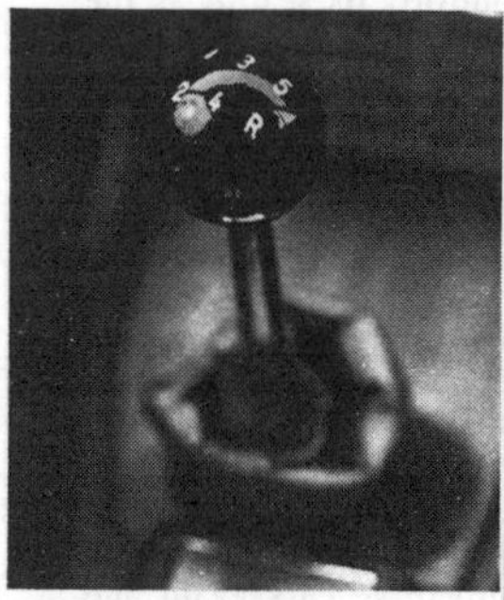

Fiat offers the only 5-speed gearbox in the group.

ENGINE & DRIVE TRAIN: 5 MEDIUM GTS

	Datsun 240Z	Fiat 124 Sport Coupe	MG B GT	Opel GT	Triumph GT6 Mk3
Engine type	L6 sohc	L4 dohc	L4 ohv	L4 sohc	L6 ohv
Bore x stroke, mm	83.0 x 73.3	80.0 x 80.0	80.3 x 89.0	93.0 x 69.8	74.7 x 76.0
Displacement, cc	2393	1608	1798	1897	1998
Bhp @ rpm	150 @ 6000	104 @ 6000	92 @ 5400	90 @ 5200	95 @ 4700
Torque @ rpm, lb-ft	148 @ 4400	94 @ 4200	110 @ 3000	111 @ 3400	117 @ 3400
Transmission	4-sp man[1]	5-sp man	4-sp man[2]	4-sp man[1]	4-sp man[2]
Standard final drive ratio	3.36:1	4.10	3.91	3.44	3.27
Engine speed @ 70 mph, rpm	3350	4000	3980	3560	3480

[1]3-sp automatic optional [2]overdrive optional

Datsun's performance is in a class by itself.

PERFORMANCE: 5 MEDIUM GTS

	Datsun 240Z	Fiat 124 Sport Coupe	MG B GT	Opel GT	Triumph GT6 Mk3
Lb/hp (test weight)	18.5	25.2	29.6	27.8	26.2
Top speed, mph	122	112	105	110	107
Standing ¼ mi, sec	17.1	18.6	19.6	18.4	18.6
Speed at end of ¼ mi, mph	84.5	72.5	72.0	74.0	74.5
0–60 mph, sec	8.7	12.4	13.6	11.9	12.0
Brake fade, % increase in pedal effort in six ½-g stops	10	nil	17	nil	20
Stopping distance from 80 mph, ft	287	319	315	277	361
Control in 80-mph panic stop	excellent	very good	fair	good	good
Overall brake rating	very good	good	good	very good	fair
Actual speed at in. 60 mph	61.5	56.0	58.0	56.0	58.0
Fuel economy, mpg (trip)	21.1	22.1	24.1	25.9	23.2
Cruising range on full tank, mi	335	260	290	340	270

THE $3500 GT

In recognition of the value of seatbelts—and in the hope that we can influence more people to use them—we're rating the belts in each car. Opel has followed U.S. GM practice by simply fitting separate lap and shoulder belts, each with its own pushbutton buckle. These fit well once in place and the roof anchorage for the shoulder belt is far superior, for instance, to the Fiat's which is on the body side behind and slightly below (that is bad—collarbones can be injured) the shoulder. But the separate belt arrangement makes it extremely inconvenient to strap oneself in, thus making it less likely a driver or passenger is going to do it.

Another safety-related item: throne-type seats used to meet the federal government's head-restraint rule impair vision to the rear, and a blind spot created by the rear roof is a further vision problem. The GT needs more outward vision for traffic maneuvering.

But all in all, it's a pleasant if not exciting little coupe. Not a bad car at all—it's just that the Datsun and Fiat are so good.

MGB GT

WE'VE HEARD that British Leyland is simply letting the MGB run its historic course; when it can't be sold anymore they'll drop it and that's that. The car seems to bear it out. Meeting the U.S. crash-safety regulation was done by laying an ugly, add-on instrument panel over the existing one and the little bit of styling facelift has been done in a haphazard way.

It's truly a car of the past. Everywhere there's evidence of a sports car designed and built in the traditional manner—in a rather homemade way, to be blunt, in great contrast to the professional design and execution of the Datsun, Fiat and Opel.

That impression carries through on the road. The GT is heavy (nearly as heavy as the Datsun) but gets only 92 bhp from its noisy pushrod engine; so it's the slowest of the group. And a rather balky shift linkage doesn't contribute to driving fun—a surprise, because this was one area in which MGs always excelled in the past.

The optional overdrive does make the B GT a capable long-distance tourer; in OD at 70 mph it's turning 3190 rpm, vs the 3980 given in the Engine & Drive Train table for normal 4th gear. And the overdrive gives it the second-best fuel economy figure. But don't expect the MG to be quiet at speed even with overdrive; there's so much wind noise you'd think it was a roadster, not a coupe.

The B handles well enough but rides very stiffly. At the limit there is a bit of oversteer that makes it fun to toss the car around, especially on low-speed curves. The steering is heavy, but the MG has the quickest steering in the group. The brakes are about average.

Ventilation, provided by a simple flap under the dash, is ineffective compared with the best of the group, but one can at least maximize it by opening the door ventwings and/or the swing-out quarter windows.

Vision outward is quite good, and MG augments it with a curious righthand fender mirror, stuck out there all by itself. The seatbelts are of the simple Kangol variety, same as on the Fiat, and someone at the factory or distributor had also installed them wrong so that one section of the belt had to be twisted.

There's little to redeem the MGB GT, not even a low price, and we can only call it a holdover from another era.

Triumph GT6 Mk 3

THE GT6 IS almost as close to the MG in our score-giving as the Fiat to the Datsun. It rates close to the Opel in performance, and with six cylinders its engine is smoother and quieter than all but the Datsun and Fiat. If overdrive is ordered, Triumph installs a 3.89:1 final drive rather than the 3.27:1 of the test car and in this form it will be a bit quicker through the gears. But what promises to be a good open-road car (if the smooth, adequately powerful engine is any indication) turns out not to be because a drumming driveline vibration sets in at about 65 mph and stays there as speed rises.

The gearbox is stiff-shifting and the shifter's H-pattern is oddly skewed; this all takes some getting used to and perhaps the owner could adapt. In any case, the GT6 is a decently enjoyable car over a curvy road at moderate speeds, with light if not particularly quick steering and good handling response. But in ultimate cornering power it is at the bottom of the group. Don't expect the brakes to accomplish much—though their fade resistance is adequate they take a very long distance to stop the car from 80 mph. On bumpy or irregular road surfaces the GT6's backbone-plus-body structure is the least staunch in the group, creaking and rattling when the going gets rough.

There are charms to the GT6. Its interior materials look the richest of any in the group and the instrument layout is particularly handsome. The seats have been upgraded in recent years (as have the MGB's), but in the GT6 they have "throne" backs like the Opel's which are a bit restrictive for rear vision. Also like the Opel, the GT6 has a blind quarter that hampers one's ability to maneuver freely in traffic.

Though the GT6 scores esthetically and in performance over the MGB, it loses it all on comfort. Its seating is the most cramped in the group, the steering wheel is very high, and its seatbelts were next to impossible to adjust to fit anybody. Triumph has gone to greater lengths to update the GT6 than MG has the B, but it still failed to make much of an impression on R&T's five testers and had to be rated last-in-group.

ONE OF OUR five drivers commented after the trip that the Datsun should be rated separately, as it is simply a class above the rest. But when all was said and tallied, the Fiat came surprisingly close and the Opel was far from unpleasant. As for the two Britishers, we do not wish to kick dead horses and sincerely hope that England will be able to get off her duff, produce some competitive cars again and challenge the other countries. We have reason to believe that British Leyland does intend to keep building sports cars and to come up not only with new designs but to realign the product "mix" of MG, Triumph and Jaguar. One of these new products, we would predict, will be a medium-price GT replacing both the B and the GT6—one that we hope will render the choice of a good $3500 GT a bit more difficult to make.

triumph GT6

INSTANT access is the GT6's forte. The whole front section folds-out allowing mechanics total freedom of action.

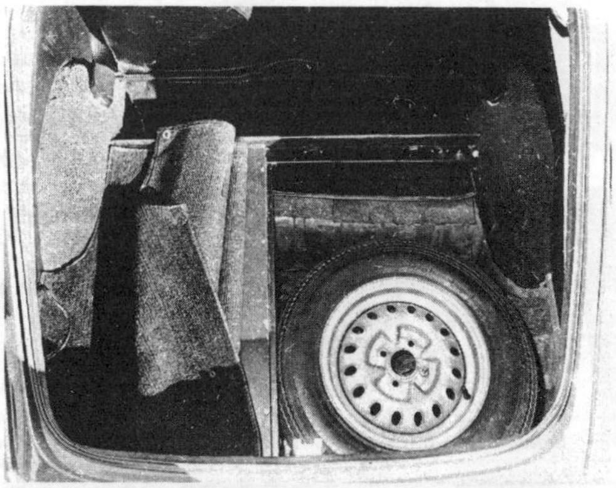

CARPET covers the spare wheel and fuel tank which nestle side-by-side under the rear platform. The shopping bag wells are at the top of the picture.

87 ►

Two axle ratios are available, 3.27 on the standard model and 3.89 on the overdrive model. This gives the standard model a higher top speed of 112 mph against the maximum of 106 mph in overdrive top. Acceleration is slightly better in the overdrive model.

The 2-litre engine is only a slightly warmed version of the unit which powers the Triumph 2000. Twin Strombergs pass the gas and 98 bhp comes out at 5300 rpm. A healthy 108 lb.ft of torque is produced at 3000 rpm.

The nett result is a quiet, smooth unit which doesn't like high revs. The GT6 isn't a stunning supercar but is a reasonably quick machine with standing quarter times in the mid-17s.

Our first drive of the car was out of the dealer's yard into the afternoon rush hour.

Our testers are at their most critical when stuck in traffic jams. Every small point seems to come out in a car as it crawls from one red light to another.

The GT6 annoyed us because of the large number of blind spots. At the point when we needed full 360 degree visibility it just wasn't there.

The very thick rear quarter panels cut a lot of vision into parallel traffic lanes, making swapping difficult. By some oversight the test car wasn't fitted with an external rear vision mirror. And for good visibility, this car could do with two exterior mirrors.

We were impressed with the ease of driving. We had no trouble dribbling the car along at less than 1000 revs in bottom gear, then booting it all the way to the red line to take advantage of a gap in traffic.

As we grew accustomed to the vehicle and sized it mentally we appreciated it more and more in traffic. Because it is so small, the car is good for city work. Its responsive steering and good take-off makes it easy to cut through traffic.

Parking too is easy. The small size and huge lock gets the car into seemingly impossible little holes.

The suspension is firm without being harsh. This firmness combined with the short wheelbase (83in — 2110mm) gives a pitchy ride, which doesn't even out with speed.

The front of the car chops up and down and the ride is distinctly uncomfortable. There is no suggestion of float and the car is perfectly under control at 100 mph. But the ride is a little too firm.

This ride stiffness carries through into cornering. The actual cornering power isn't very high in comparison with some sports cars, but control is very high.

The initial steering reaction is understeer. This persists all the way around a corner unless a lot of power is applied to poke the tail out.

Very rapid motoring is likely to be accompanied by alternate dramatic under and oversteer and should be reserved for the skilful.

Our final and most profound impression of the GT6 was its age. It is an old car in the traditional English pattern.

The dashboard is real wood, the ride hard and the handling at speed for experts only. It is quiet and the engine is smooth.

But it is not much fun being blasted-off by Falcon GTs, Monaro GTSs and their like — even if you are in a "genuine" sports car. The price of purism comes high. ■

COCKPIT treatment is almost pure racer. The tight-fitting seats and body are well padded in the right places to provide support and comfort.

AUTO TEST

TRIUMPH GT6 Mk III

Close-fitting fun car

AT-A-GLANCE: Latest version of Triumph's small but big-engined sports coupé. Mechanical specification and performance unchanged from Mk II but body changes have tidied-up styling and improved refinement. Fast and safe-handling, braking only fair. Economical and competitively priced.

THE latest versions of the Triumph GT6 and Spitfire were announced just after last year's London Motor Show, and like the newest variations on their MG counterparts were a "freshening up" exercise on well-tried designs rather than anything really new. Important from Triumph's point of view is that their new styling gives them a family similarity to the rest of the model range. Few would dispute that the changes have improved the appearance of the cars, the GT6's particularly so. What was always an attractive idea spoiled outwardly by fussy and badly planned detail has become a much smoother, more coherent design. The removal of the wing-top raised seams on the one-piece bonnet section is particularly welcome, while a new Stag-type undercut tail with new rear lamp clusters and a full-width bumper improve the look of the rear. Other exterior changes are a slightly deeper windscreen, a matt black grille surrounded by thick polycarbonate over-riders below the front bumper, flared wheel arches, new style disc wheels and a reshaped engine hump on the bonnet which no longer has unsightly louvres on its top and sides. The GT6 Mk III also incorporates a number of interior changes.

The GT6 was introduced in 1966. Its small, low coupé body is directly derived from that of the Triumph Spitfires which raced at Le Mans (and were timed at 130 mph) in 1964. At first it was sold almost entirely for export, though at £1,000 it was very reasonably priced; the ways of the financial world have added £250 to the price since then but the Mk III is still good value comparatively. The first GT6 was fiercely criticized for the inadequacy of its rear suspension, which was pure Herald swing-axle, coping with twice the horsepower. A much-improved Mk II version was brought out in autumn 1968 and this mechanical specification remains virtually unchanged for the Mk III. The essence of the rear suspension improvement is the use of a transverse leaf spring as an upper wishbone, and a reversed lower wishbone with semi-trailing arm providing positive location. The drive-shafts are joined by rubber "doughnut" couplings at their outer ends.

The result, as we reported in the *Autotest* of the Mk II (*Autocar* 3 April 1969) is that the car can now be cornered with confidence at speeds which would have been positively dangerous in the early GT6. Straight-line stability has improved, there is less attitude change on acceleration and braking and the car has much greater "swervability"—response to an emergency situation—than before.

With a front weight bias, the latest GT6 shows a considerable amount of understeer, with the tyres scuffing on the approach to full lock with the power still on. Eventually the inside rear wheel lifts and spins, but when traction is broken there is no sudden dramatic reaction. Backing off the throttle in mid-corner causes the front to tuck-in but not in a vicious manner. In the wet its performance is similarly predictable and adhesion on the test car, shod with the standard Dunlop SP68 radials, was good. In all, the handling is very safe and remarkably good for an arrangement which is still simple and rather unsophisticated in its operating principles.

The steering is very light and "quick", allowing the car to be accurately placed on corners. It has 4.2 turns from lock to lock, but is not noticeably low geared when one takes into account the very tight turning circle typical of all Herald derivatives. Turning on full lock, when it requires only a little over 25ft. between kerbs, the steering geometry is forced into angles which are far from ideal and is accompanied by vicious tyre scrub even when this is done at very low speed.

The suspension gives a slightly choppy and lively ride but is not *too* firm and considerably better than the lighter Spitfire (which has a simpler rear suspension layout). The jolting experienced over bad surfaces is accompanied by some noise from the rear suspension itself but no shake from the scuttle or bonnet mountings like some cars of this type. On moderately smooth roads this ride proves a good compromise between the optimum for handling and the requirements for comfort.

Unlike the Mk II that we tested, this latest car came to us without overdrive. The overdrive car has a 3.89 to 1 final drive ratio which gives a very similar overall ratio in overdrive top to the 20.15 mph per 1,000 rpm that this car has in direct top with the standard 3.27 to 1 axle. Our figures show that the performance for the two cars is very similar (the lower axle gives the overdrive car slightly better acceleration lower down the speed scale; slower to 90 and 100 mph) and that the fuel consumption is slightly better in the non-overdrive car. The conclusion from this is that, from a performance point of view, the overdrive does not represent a worthwhile investment for this model if the 3.89 axle is used. Fitted in conjunction with the 3.27 ratio touring fuel consumption would undoubtedly be improved (though to the detriment of the acceleration figures) but this is not available as a factory-fitted option. As it is, an overall figure of 27.6 mpg for the test Mk III, with a figure of 31.8 over 500 miles of mainly fast motoring, must be considered very good indeed.

This car was in fact 5 mph faster in maximum speed than the overdrive Mk II,

corresponding to 5,550 rpm, which is 50 rpm into the yellow "caution" areas on the tachometer; the red line is at 6,000. The legal limit calls for a restful 3,400 rpm. Without overdrive, third gear has a long-legged maximum of 96 mph which makes it useful for overtaking under fast main road conditions.

The 98 bhp engine has the sort of smooth, lusty performance that one would expect from a 2-litre six-cylinder. Carburation is by twin Stromberg-CD side-draught carburettors. It proved an untemperamental starter, with choke needed for only a short period from cold. Its usually even 750 rpm tickover became less so as the test went on, and turned into a tendency to stall at the end of our period of tenure. This was simply cured by topping up the carburettor dashpots. The exhaust noise is not very "sporting", despite its twin tail pipes sprouting from a TR6-like transverse silencer, but neither is it obtrusive.

The gearchange has short movements and a positive, though notchy, engagement. The clutch is light, requiring an effort of only 25lb, and smooth in action.

Initial braking "feel" was not very confidence-inspiring, and a maximum retardation of 0.92g before the front wheels locked in our braking tests rates as only moderately good. At lower pedal pressures their performance was, in fact, proportionately better than on the Mk II and the early locking of the front wheels suggests that the rear drums were in need of some adjustment. This is further suggested by the poor performance of the handbrake which would not hold the car on a slope steeper than 1 in 6, and even then only with the force of both hands to pull the lever on to the last tooth of the ratchet. The car had no difficulty in restarting on gradients up to 1 in 3. The brakes performed well in our fade tests, with no increase in pedal pressure after 10 stops from 70 mph.

The interior changes which accompanied the restyling included new seats and a flat alloy, leather rimmed steering wheel. During the life of the Mk II the rear luggage floor was cut away to allow reclining seats; a great benefit to taller drivers who can now find whatever driving position they require, thanks to the recliners and greater rearward seat adjustment. He (the taller man) also appreciates the 2in. greater depth in the windscreen, though the downward sweep of the roof to meet the side windows, and the thick rear quarters framing the steeply raked rear window, mean that the car still does not score many points for all-round visibility. External mirrors would be a help. Conversely, the shorter driver may find the seats too low. Some find the interior of this close-couple coupé rather claustrophobic, though the black headlining which used to accentuate this effect has now been replaced by a light-coloured one.

Subtle restyling has given the GT6 a neater, more modern appearance. Lift-up rear door (above) gives wide access to luggage area, though sill is rather high. The Mk III has new pressed-steel wheels (opposite, below) without hub caps; metal centre-trims are held on by wheel nuts. The lift-up front body section gives excellent accessibility to engine compartment and front suspension (opposite)

Having a tight-fitting cockpit has advantages and disadvantages. It helps the already good seats to provide very positive sideways location, effectively extending their support up to the sides of the doors and the padded transmission tunnel and handbrake cover-cum-armrest; the padding also extends to the gearbox housing. On the other hand, it means that the seat recliners and adjusters are almost impossible to operate with the doors closed and that the standard-fitment Britax Twin-Lok seat belts, which are so good in some cars, are difficult to adjust and uncomfortable to wear. It also means that it is difficult to place things in the back of the car except through the spring-loaded lift-up rear door /window.

Reshaping the rear floor to allow more seat adjustment has produced a useful well behind the seats where items can be stowed securely

TRIUMPH GT6 Mk III (1,998 c.c.)

ACCELERATION

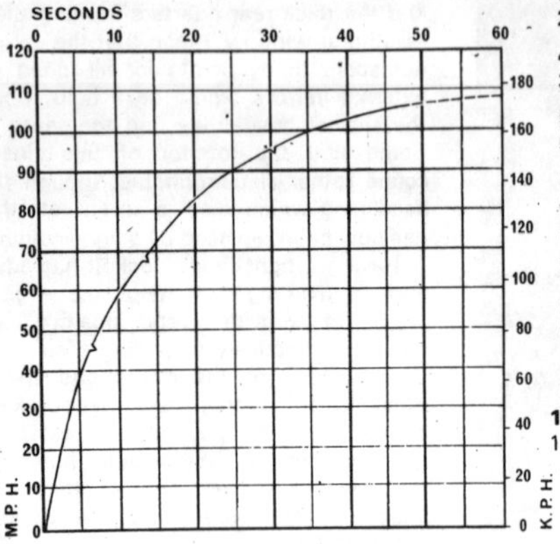

SPEED MPH TRUE	INDICATED	TIME IN SECS
30	29	3.9
40	39	5.5
50	50	7.7
60	60	10.1
70	70	14.0
80	80	18.4
90	90	24.5
100	100	35.8

GEAR RATIOS AND TIME IN SEC

mph	Top (3.27)	3rd (4.11)	2nd (5.82)
10-30	—	7.1	5.2
20-40	8.6	6.7	4.5
30-50	7.8	6.6	4.5
40-60	8.3	6.5	4.7
50-70	9.2	7.1	—
60-80	10.1	8.5	—
70-90	11.7	11.0	—
80-100	16.9	—	—

Standing ¼-mile
17.4 sec 78 mph
Standing Kilometre
32.5 sec 97 mph
Test distance 1,497 miles
Mileage recorder
1.3 per cent over-reading

PERFORMANCE

MAXIMUM SPEEDS

Gear	mph	kph	rpm
Top (mean)	112	180	5,600
(best)	112	180	5,600
3rd	96	155	6,000
2nd	68	110	6,000
1st	46	73	6,000

BRAKES

FADE
(from 70 mph in neutral)
Pedal load for 0.5g stops in lb

1	**42**	**6**	**40-35**
2	**45**	**7**	**35**
3	**50-45**	**8**	**35**
4	**45-40**	**9**	**40-35**
5	**40-35**	**10**	**40-35**

RESPONSE (from 30 mph in neutral)

Load	g	Distance
20lb	0.18	167ft
40lb	0.46	65ft
60lb	0.67	45ft
80lb	0.92	33ft
90lb	0.90	33.4ft
Handbrake	0.24	125ft

Max. Gradient 1 in 6

CLUTCH

Pedal 25lb and 5in.

COMPARISONS

MAXIMUM SPEED MPH

Triumph GT6 MkIII	**(£1,254)**	**112**
Ford Capri 2000GT	(£1,242)	106
Opel Manta 1.9	(£1,476)	105
Toyota Celica 1600ST	(£1,362)	105
MGB GT	(£1,389)	102

0-60 MPH, SEC

Triumph GT6 MkIII	**10.1**
Ford Capri 2000GT	10.6
Toyota Celica 1600ST	11.5
Opel Manta 1.9	12.2
MGB GT	13.0

STANDING ¼-MILE, SEC

Triumph GT6 MkIII	**17.4**
Toyota Celica 1600ST	18.2
Opel Manta 1.9	18.2
Ford Capri 2000GT	18.2
MGB GT	18.5

OVERALL MPG

Toyota Celica 1600ST	28.0
Triumph GT6 MkIII	**27.6**
Opel Manta 1.9	25.3
MGB GT	23.7
Ford Capri 2000GT	22.0

GEARING

(with 155SR13in. tyres)

Top	20.15 mph per 1,000 rpm
3rd	16.0 mph per 1,000 rpm
2nd	11.29 mph per 1,000 rpm
1st	7.59 mph per 1,000 rpm

CONSUMPTION

FUEL
(At constant speed—mpg)

30 mph	47.4
40 mph	47.5
50 mph	45.5
60 mph	41.7
70 mph	33.3
80 mph	29.2
90 mph	24.8
100 mph	21.5

Typical mpg 27.8 (10.1 litres/100km)
Calculated (DIN) mpg 30.3 (9.3 litres/100km)
Overall mpg 27.6 (10.2 litres/100km)
Grade of fuel Super, 5-star (min. 99.4 RM)

OIL
Consumption ... Negligible

TEST CONDITIONS:
Weather: Dry, dull. Wind: 5-10 mph.
Temperature: 19 deg. C. (66 deg. F.)
Barometer: 29.6in. hg.
Humidity: 75 per cent.
Surfaces: Dry concrete and asphalt.

WEIGHT:
Kerb weight 18.06 cwt (2023 lb–918 kg).
(with oil, water and half full fuel tank)
Distribution, per cent F, 54.6; R, 44.4
Laden as tested: 22.1 cwt (2,473 lb–1,123 kg).

TURNING CIRCLES:
Between kerbs L, 25 ft 1 in.; R, 25 ft 9 in.
Between walls L, 26 ft 11 in.; R, 27 ft 6 in.
Steering wheel turns, lock to lock 4.2.
Figures taken at 12,500 miles by our own staff at the Motor Industry Research Association proving ground at Nuneaton.

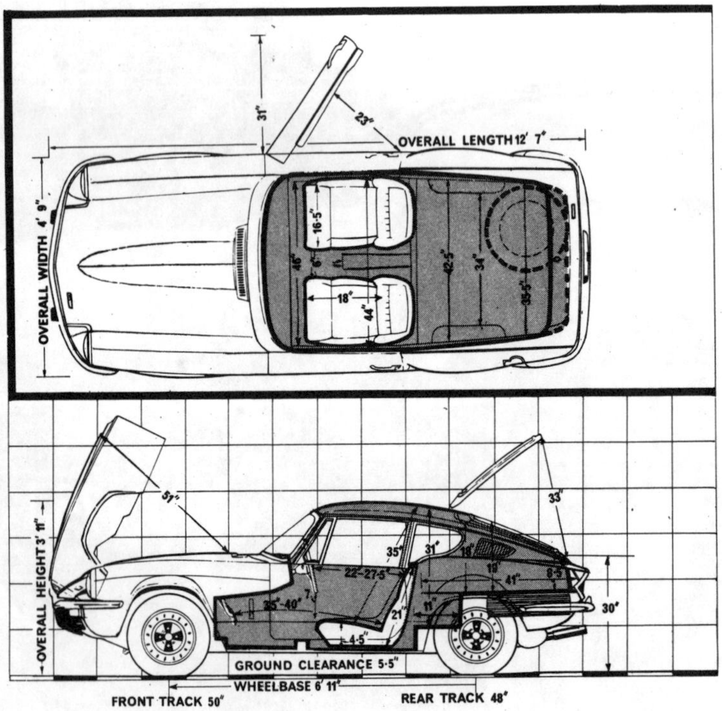

STANDARD GARAGE 16ft x 8ft 6in.

;PECIFICATION

RONT ENGINE, RONT-WHEEL DRIVE

NGINE	
ylinders	6, in line
1ain bearings	4
ooling system	Water, sealed system: pump, fan and thermostat
ore	74.7mm (2.94in.)
troke	76.0mm (2.99in.)
isplacement	1,998 c.c. (122 cu.in.)
alve gear	Overhead, pushrods and rockers
ompression ratio	9.25-to-1 Min. octane rating: 99.4RM
arburettors	Twin Stromberg 150CD
uel pump	Mechanical diaphragm type
il filter	Full flow, renewable element
1ax. power	98 bhp (net) at 5,300 rpm
1ax. torque	108.3 lb.ft (net) at 3,000 rpm
RANSMISSION	
lutch	Borg and Beck, diaphragm spring, 8.5in. dia
earbox	Four-speed, all-synchromesh
ear ratios	Top 1.0
	Third 1.25
	Second 1.78
	First 2.65
	Reverse 3.10
nal drive	Hypoid bevel, 3.27-to-1, optional 3.89-to-1 with overdrive
HASSIS and BODY	
onstruction	Separate cruciform chassis
USPENSION	
ont	Independent, coil springs, wishbones and telescopic dampers
ear	Independent, transverse leaf spring, lower links, semi-trailing arms, telescopic dampers
TEERING	
pe	Alford and Adler, rack and pinion
heel dia	15in.
RAKES	
ake and type	Girling disc front, drum rear
rvo	None
mensions	F 9.7in. dia.
	R 8.0in. dia. 1.25in. wide shoes
vept area	F 197 sq.in., R 63 sq.in.
	Total 260 sq.in. (245 sq.in./ton laden)
HEELS	
pe	Pressed steel disc, ventilated
	4½in. wide rim
res—make	Dunlop SP68
—type	radial ply tubeless
—size	155SR13in.
UIPMENT	
ttery	12 Volt 56 Ah.
ternator	28 amp a.c.
adlamps	Lucas sealed beam 120/90 watt (total)
versing lamp	2 standard
ectric fuses	3
reen wipers	Two-speed
reen washer	Standard, manual plunger
terior heater	Standard, fresh air
ated backlight	Standard
fety belts	Standard Britax Twin-Lok
terior trim	PVC seats, PVC headlining
oor covering	Carpet
ck	Screw scissors
cking points	4, under frame
indscreen	Zone toughened
derbody protection	Phosphate dipping before painting
AINTENANCE	
el tank	9.75 Imp. gallons (44.3 litres)
oling system	11 pints (including heater)
gine sump	8 pints (4.5 litres) SAE10W/40. Change oil every 6,000 miles. Change filter element every 12,000 miles.
arbox	1½ pints SAE90EP. Change oil every 6,000 miles
al drive	1 pint SAE90EP. Change oil every 6,000 miles.
ase	2 points every 12,000 miles
e pressures	F 24; R 28 psi (all conditions)
x. payload	448 lb. (203 kg.)
RFORMANCE DATA	
gear mph per 1,000 rpm	20.15
an piston speed at max. power	2,650 ft/min.
p per ton laden	88.9

AUTOTEST TRIUMPH GT6

and relatively out of view; otherwise there is an open parcels shelf on either side of the facia, though the radio loudspeaker occupies some of the space on the passenger side. The spare wheel, jack and four-piece tool kit are housed under the rear floor, which, like the remainder of the interior, is covered in neat black carpeting.

The latest dashboard design (like that of the Mk II) is well planned. The facia is covered in wood treated with a non-reflective coating which looks and wears well. All instruments have neat white-on-black dials. The light switches are of rocker type with the two-speed wipers and electric washers operated by a square knob on the right; for some reason the test car had the wipers arranged in the left-hand drive pattern with the maximum swept area on the left of the screen. Indicators and dipping are looked after by a column stalk. A four-way hazard warning device is fitted.

A point which deserves criticism is the new steering/starter lock which, in common with similar fitments on other British Leyland sports cars, seems to have been situated with a view to convenience of installation rather than of use. It is hidden well below the dashboard in an awkward housing, difficult to reach (impossible if strapped in), not illuminated and difficult to engage with the key.

The heater controls are much better than they were on the first GT6, with proper slides, and a pull-out knob built into one for the single-speed fan. The heater itself is still of the water valve type. Criticism in earlier cars of heat entering the cockpit from the engine compartment is no longer valid, though on a hot day the interior can get stuffy, despite the flow-through ventilation system now fitted. This has two eyeball ventilators on the front of the dashboard and two beneath it; the air pressure is not strong (the pipes are of narrow diameter) and one has to choose between air to the face or air to the foot wells and operate the eyeball adjusters accordingly. There are also conventional opening front quarter lights and the rear side windows can be opened to assist extraction of stale air. Misting over of the rear window can be a problem in a car with such a small interior area and a heated rear window is fitted as standard equipment on the GT6.

In common with all the Herald range (of which the GT6 and the Spitfire are now the only survivors) accessibility for maintenance is unrivalled, with the whole of the bonnet section including the front wings hinging forward out of the way.

The GT6 is an uncompromising two-seater with very good performance. Its small size gives it a lot of its appeal but also creates its limitations, for the interior space it provides is smaller than that of most of the cars with which it competes. But it has been considerably improved over the years, and now has a chassis and fitments complementary to an always-pleasant power unit, and this refinement has come without increasing the price out of reasonable bounds. □

Interior is well planned but a tight fit. Scope for greater seat adjustment than before improves driving position. Below: spare wheel and tool kit are housed beneath the rear floor

MANUFACTURER:

Triumph Motor Company Ltd, Coventry

PRICES	
Basic	£1,002.00
Purchase Tax	£252.38
Total (in G.B.)	£1,254.38
EXTRAS (inc. P.T.)	
Overdrive	£68.75
Wire wheels	£47.50
Laminated windscreen	£8.75
PRICE AS TESTED	£1,254.38

76
GT6
CIK-876

GT6 IS FUN, BUT IS THAT ENOUGH ON...

TRIUMPH'S THIRD TIME ROUND

WHEELS ROAD TEST

TRIUMPH'S GT6 MARK 3 is ideal for the enthusiast who wants to recline at full arm's length from its leather-bound wheel and drive with the seat of his pants.

It's that sort of car — a taut, stubby little sportster you grab and fling around just for the sheer fun of it.

But is this the answer in a market becoming used to the sophistication of coupes like the Datsun 240Z which handles even more capably than the GT6, goes

ABOVE LEFT: The rakish lines have not changed appreciably throughout the GT6's five years of production. Higher bumpers have been added to comply with American safety regulations.

BELOW LEFT: With all its apertures open, the car loses its stylish lines. The huge bonnet lifts to expose completely the engine and the heated rear window opens for access to the luggage compartment.

PHOTOGRAPHY: UWE KUESSNER

A rejuvenated GT6 comes over well until you begin to compare it with the strong opposition from Fiat and Datsun. Given a price tag $500 lower it would begin to make sense, reports David Varley.

much harder and has more refinement in its cockpit?

You see, the GT6 is really a hybrid — it didn't even get much of a birthright to boast about. The Triumph midwives simply dropped their 2000 cc sedan engine into a Spitfire chassis and covered it all with a solid top.

In its five years of production the GT6 has undergone substantial change. The major problem of shocking handling was overcome in the Mark 2 model when the Herald-style, rear swing axles were dropped and a semi-trailing arm system introduced. With the original system in the Mark 1 the rear wheels tucked under, making the car one of the world's worst desperate oversteers.

With the Mark 3 the major alterations are in styling. It has joined its bigger brothers with the typical Triumph flat recessed rear panel bordered by stainless steel strips. To meet American safety standards it, too, has been given higher bumpers and a slightly revised front end.

Filling the power bulge on the long, drooping bonnet, which spears its way forward from the windscreen, is the same 1998 cc unit Triumph normally housed in the basic 2000 sedan.

In the lighter, more streamlined body of the GT6 it has the chance to really prove itself and does so in a way that makes you think the car and engine had originally been designed as a team.

Performance, through the four-speed gearbox (with overdrive on third and fourth) is on a par with the average sedan but not in the class of the big GTs like the Charger or Falcon.

However, the little six is happy to dawdle along at low revs around town or to fly up to its red-line maximum of 6000 rpm whenever needed. It shows little complaint when asked to pull away from 20 mph in top gear and it is still singing happily when a heavy right foot takes it over the three figure mark.

The gearbox is a major disappointment. Although the four ratios are right for most tastes, the change itself is strange. Sometimes it will glide through the gears with silky smooth precision, others it baulks and refuses to go in. During acceleration tests it constantly grated between first and second and occasionally offered the same resistance going into third.

The gearlever is mounted almost horizontal in the transmission tunnel exactly where the driver's left hand drops from the wheel. Its movements are short and could be rapid if it wasn't for the rather slow syncromesh. The knob on top of the lever is large, and houses a small switch for thumb operation of the overdrive. This idea is terrific in every aspect except one — it wears out the left thumb of your driving gloves!

TOP: On the Mark 3, the coupe styling melts into the now traditional Triumph recessed rear panel with its stainless steel surround.

The cockpit, although tiny, is very comfortable offering a full arms' length driving position for most people. The ignition key, which is impossible to reach when belted in, can be seen under the dash.

The optional overdrive is advisable in the GT6 if you are intending doing quite a few high speed trips. Not only does it cut down the revs in both third and top gears but reduces noise and offers better fuel economy. Cruising in the region of 90 mph is effortless in fourth overdrive with the tachometer showing only a moderate 4750 rpm.

Driving through the suburbs, third overdrive was easier to use than direct top. These gears have very similar ratios (3rd o/d is 3.91 while fourth is 3.89) and third offers instant acceleration with just the flick of the gearstick mounted switch.

With the optional overdrive the final drive ratio is 3.89 for 21.14 mph per/1000 rpm while the standard diff ratio of 3.27 gives 20.1.

The only fault we found with the overdrive was its delay in engaging. It took a couple of seconds to operate after switching.

In the middle speeds, changing into overdrive produces a severe jolt bringing complaints from passengers about the driver's skill. The only way to get rid of this is to time the change and blip the clutch at the right time. This can be difficult with the delay in engaging the overdrive.

Under acceleration the overdrive disengages instantly at the flick of the switch.

The clutch pedal is very light with a medium travel of about five inches. The clutch itself did not prove man enough for our acceleration tests and slipped momentarily before the car got underway. With the car's high ratio first gear we found it necessary to rev it to about 5000 rpm before releasing the clutch and getting underway. At lower revs the car just bogged down completely for a precious fraction of a second before heading down the strip.

Half-way through our acceleration runs the engine started misfiring badly.

(Continued on page 80)

WHEELS ROAD TEST

TECHNICAL DETAILS

TRIUMPH GT6 MK3

MAKE . . . **Triumph**
MODEL . . . **GT6 MK3**
BODY TYPE . . . **Coupe**
PRICE . . . **$4163**
OPTIONS . . . **Overdrive**
COLOR . . . **Orange**
MILEAGE START . . . **3266**
MILEAGE FINISH . . . **4027**
WEIGHT . . . **2089 lbs (453 kg)**

FUEL CONSUMPTION:
Overall . . . (8.8 kpl) 25 mpg
Cruising . . . (8.5 kpl-9.9 kpl) 24-28 mpg

TEST CONDITIONS:
Weather . . . cool, cloudy
Surface . . . Castlereagh Drag Strip
Load . . . two persons
Fuel . . . premium

SPEEDOMETER ERROR (mph):

Indicated	30	40	50	60	70	80	90
Actual	28	37	47	57	66	75	85

PERFORMANCE

Piston speed at max bhp . . . (805.6 m/min) 2640 ft/min
Top gear mph per 1000 rpm . . . 20.1 (36.1 kpl)
Engine rpm at max speed . . . 5000

MAXIMUM SPEEDS:
Fastest run . . . (172 kph) 108 mph
Average of all runs . . . (169 kph) 106 mph
. . . Speedometer indication, fastest run (184 kph) 115 mph

IN GEARS:
1st . . . 42 (67 kph) (6000 rpm)
2nd . . . 51 (82 kph) (6000 rpm)
3rd . . . 86 (137 kph) (6000 rpm)
3rd o/d . . . 101 (161 kph) (6000 rpm)
4th . . . 104 (166 kph) (5500 rpm)
4th o/d . . . 106 (169 kph) (5000 rpm)

ACCELERATION (through gears):
0-30 mph . . . 3.3 sec
0-40 mph . . . 5.5 sec
0-50 mph . . . 7.8 sec
0-60 mph . . . 10.1 sec

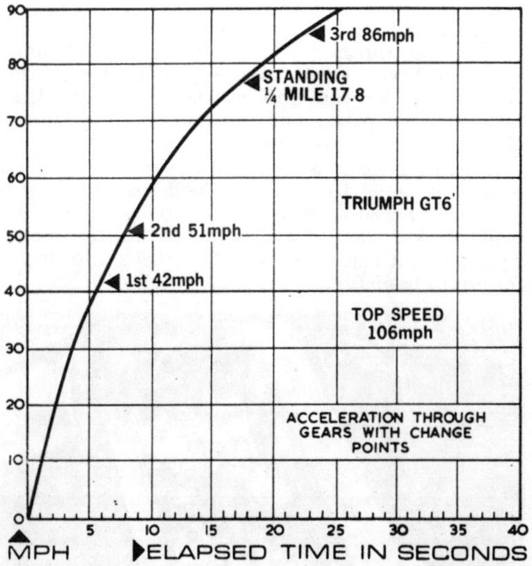

0-70 mph . . . 14.0 sec
0-80 mph . . . 19.4 sec
0-90 mph . . . 25.1 sec

	2nd gear	3rd gear	4th gear
20-40 mph	3.7 sec	5.0 sec	6.7 sec
30-50 mph	3.7 sec	5.6 sec	6.4 sec
40-60 mph	–	5.4 sec	7.2 sec
50-70 mph	–	6.2 sec	7.2 sec

STANDING QUARTER MILE:
Fastest run . . . 17.7 sec
Average all runs . . . 17.8 sec

BRAKING:
From 60 mph to 0 . . . 3.6 sec

SPECIFICATIONS

ENGINE:
Cylinders . . . six in line
Bore and stroke . . . 74.7 mm (2.94 in.) x 76 mm (2.99 in.)
Cubic capacity . . . 1998 cc (122 cu in.)
Compression ratio . . . 9.25 to 1
Valves . . . pushrod ohv
Carburettor . . . 2 Stromberg 1.5 CD
Fuel pump . . . mechanical
Oil filter . . . full flow
Power at rpm . . . 98 bhp at 5300 rpm
Torque at rpm . . . 108 lb/ft at 3000 rpm

TRANSMISSION:
Type . . . four speed all syncro with overdrive
Clutch . . . single dry plate diaphragm
Gear lever location . . . centre floor

RATIOS:

	Direct	Overall	mph per 1000 rpm	(kph)
1st	2.07	10.29	6.05	(9.6)
2nd	1.77	6.92	9.50	(15.2)
3rd	1.25	4.88	13.51	(21.73)
3rd o/d	1.005	3.91	16.66	(26.3)
4th	1.00	3.89	16.89	(27.1)
4th o/d	0.77	3.12	21.15	(36.1)
Final drive	3.89			

CHASSIS AND RUNNING GEAR:
Construction . . . separate chassis with bolted-on body
Suspension front . . . independent by wishbones and coil springs with anti-roll bar
Suspension rear . . . independent with transverse upper leaf spring, lower wishbones and radius arm
Shock absorbers . . . telescopic
Steering type . . . rack and pinion
Turns l to l . . . 4.5
Turning circle . . . 25 ft 3 in. (7.7 m)
Steering wheel diameter . . . 15 in.
Brakes type . . . disc/drum
Dimensions . . . 9.7 in. discs, 8 in. drums
Friction area . . . 260 sq in. (1676 sq cm)

DIMENSIONS:
Wheelbase . . . 84 in. (2133 cm)
Track front . . . 4 ft 1 in. (1244 cm)
Track rear . . . 4 ft 1 in. (1244 cm)
Length . . . 12 ft 4 in. (3784 cm)
Width . . . 4 ft 10 in. (1473 cm)
Height . . . 3 ft 11½ in. (1205 cm)
Fuel tank capacity . . . 9¾ gals (44.3 litres)

TYRES:
Size . . . 155SR-13
Pressures . . . F 24 (1.66 kg/cm^2) R 30 (2.1 kg/cm^2)
Make on test car . . . Dunlop SP 68

GROUND CLEARANCE:
Registered . . . 4 in. (102 cm)

TRIUMPH'S THIRD TIME ROUND

Continued from page 78

This was mysteriously cured simply by taking the car out for a fast run. It played up a couple of times later but was cleared each time by a run at high revs for a few hundred yards.

Since the suspension specialists got their fingers onto the Mark 1, the GT6's handling has improved considerably but it is still not up to the standards expected from such a sporting car.

With its trailing arm rear suspension the car understeers strongly when thrown into a bend under power. This makes the driver haul on more and more lock in a most unsportscar-like manner.

But let off the power or worse still, brake in a corner and the handling of the Mark 3 changes to severe oversteer. To correct this the driver has to be very fast in pouring on opposite lock. With over four turns of the wheel from lock to lock, this is quite a feat.

So the everyday cornering capability of the GT6 is still no more than that of many quick sedans.

Nevertheless, a good driver can learn to use the oversteer and apply it to good use. But learning is a twitchy business.

The car has a firm ride — fine for highway cruising, but choppy in the rough. Typically sportscar hard, the suspension allows the car to walk around on only slightly rugged surfaces with the tail doing the military two-step around the bitumen.

No one will find trouble parking a GT6. Its steering lock is so good the front wheels almost turn at right angles, much the same as the old Triumph Herald. On full lock the engine is really straining to push the car forward, almost pushing the wheels sideways. This — combined with light steering at slow speeds — makes it a joy to park in spaces previously used exclusively by Minis.

The brakes feel soft but pull the car up well in a straight line when really forced on. The test car's braking was marred by each wheel locking-up independently but this didn't put the car off line.

The handbrake is poor. Unless the driver exerts his full might to pull the high, centre-mounted lever up it will not hold the car.

One of our main criticisms of the car is its wind noise. At high speed the pleasant engine noise disappears and is taken over by howling wind, drumming into both the driver and passenger's ears. The main cause of the noise is the bad fitting around the tops of the rimless glass in the doors and the rain guttering running along the top of the windscreen. The usual method of rectifying this, by opening the windows slightly, makes no difference.

In an automatic car wash the passenger was showered, soaped, washed and ready to go out for the night by the time the process ended. Water pours in around the top of the windows, through the surrounds of the back window and from under the dashboard. The next day the driver's clutch foot was soaked as water dripped from an unknown source under the dash.

For the Australian climate the GT6 ventilation is hopeless. It might be fine in chilly England but the eye level vents in the dash let in only minimal air supplies while those under the dash don't seem to work at all. The only way to get even slightly cooled in the hot black interior is to open all the windows, including the rear quarter vents.

The car is fitted with a heated rear window as standard.

The little bucket seats are comfortable, with terrific lateral support. They are adjustable in both rake and fore and aft movement and cater for almost everyone. We had only one complaint from a lanky six footer who said he couldn't get far enough back to get a full arms' length grip on the wheel. The seats could have had more aft movement without reducing the luggage space to any great extent. They are also very close to the doors for there is little shoulder room.

The only problem behind the wheel are the pedals which are offset to the right. The accelerator is too high, making it impossible to toe and heel. On top of this there is no room for the driver's left foot except under the clutch and he has to bend his knee and rest it against the padded transmission tunnel to avoid the steering wheel.

Behind the two seats there is only cramped luggage space although this easily reached through the opening rear window. Our photographer put his heavy camera case in the back at one stage only to have it clout the driver on the back of the neck every time he applied the brakes. It is advisable to pack everything tightly to avoid a nasty accident.

Under the luggage compartment, and very hard to get out is the spare tyre and small tool kit.

For the do-it-yourself mechanic, the GT6 is a delight. The forward-hinged bonnet and wings leave the engine sitting almost naked between the front wheels. It is one of the easiest serviced engines ever. The regularly looked at features are within easy reach of even the amateur mechanic.

The dashboard is attractively finished in non-reflective wood with two big dials partially obscured by the large steering wheel. Two other smaller ones in the centre of the dash are the fuel and temperature gauges. The two big ones house the speedometer, odometer and trip meter with the tachometer in the other. In such a sports car the lack of an oil pressure gauge and an ammeter is disappointing.

The switches are an odd mixture. In front of the driver on the right is a large knob for the two-speed wipers and washers, on the left the same sort of knob for the choke (which, incidentally, was never needed on the coldest of Sydney mornings). In the centre is a rocker switch for the hazard warning lights. Mounted below the heater controls over the transmission tunnel are rocker switches for the rear window demister, interior light and headlights.

Several controls are particularly difficult to operate — the door catches, set low-down near the floor at arm's length, the window winders which the driver can't reach when belted in, the ventilators under the dashboard and worst of all the combined ignition and steering lock. This is impossible to reach with the safety belt on and at best is fiddly even without the belt.

To make things even more complicated Triumph insists on having three separate keys for the car. One, single-sided, for the ignition, one double-sided, for the doors and petrol cap, and the other (also double sided) for the rear window.

With its price of $4163 (with overdrive) the Mark 3 GT6 is going to find it a hard road to drive in Australia, because, when its all boiled down, it's really only a fun car in the Midget Spitfire mould. This makes it enjoyable, sure, but that's a lot to pay for pretty basic fun. *

The engine is one of the most easily serviced in any car. With both front wings and bonnet tilting forward, the engine sits almost naked between the front wheels.

BRIEF TEST

TRIUMPH GT6 MK 3

The GT6 has been with us for five years and in that time it has undergone considerable change. The most significant improvement appeared with the Mk 2, when the Herald-inspired swing axles were replaced by a semi-trailing arm system. Also on the Mk 2 was an uprated engine, which provided a useful increase in performance for very little sacrifice in economy.

The Mk 3 appeared in October 1970, with many detailed refinements including perforated disc wheels, flared arches on the front wings, the elimination of the bonnet louvres as well as a reduction in the size of the "power bulge", a windscreen deepened by 2in, anti-lift wipers, and flush fitting door handles with anti-burst locks. The ventilation was also improved by enlarging the intakes and outlets. The most obvious alteration was to the tail, though, which inherited the now characteristic Triumph recessed panel bordered by stainless steel strips, like that on the Stag.

The net bhp of the 1998 cc engine is now quoted as 98 bhp instead of 104. This is not due to detuning, but to a revision of test conditions in the laboratory. The output is in fact unchanged and gave a top speed of 110.5 mph. Our test car did not have overdrive and therefore sported the higher 3.27:1 axle so the resultant acceleration figures are thus slightly poorer than those we recorded for the Mk 2 that *did* have overdrive. Even so the performance is still very respectable, 0-60 mph taking 10.0 sec and the quarter mile coming up in 17.6 sec.

Top gear acceleration is also a bit poorer (inevitably) with the higher axle ratio, 20-40 mph taking 8.6 sec instead of 6.6 sec. and 70-90 mph taking 11.8 sec. instead of 9.7 sec. The maximum gear speeds indicate why it is necessary to make full use of the gearbox for rapid cross-country journeys. First gear gives 46 mph, second 68 mph and third an incredible 96 mph. But despite the high gearing the engine, which is doing around 5000 rpm at 100 mph in top, gets rather fussed at speeds over 90 mph, and does not therefore encourage high speed cruising.

Fuel consumption is where the GT6 scores over most of its rivals. At 40 mph it is doing almost 40 mpg and despite a lot of high-speed driving we managed a remarkable overall consumption of 30.5 mpg.

Top: the Mk3 is easily distinguished by its new wrap-round bumpers and the characteristic Triumph recessed panel at the rear. Above: the heated rear window opens to give good access to the luggage platform. Right: the reclining seats are comfortable and offer good support, but legroom is lacking. Below: the combined ignition and steering lock cannot be reached when wearing a seatbelt if the seat is right back

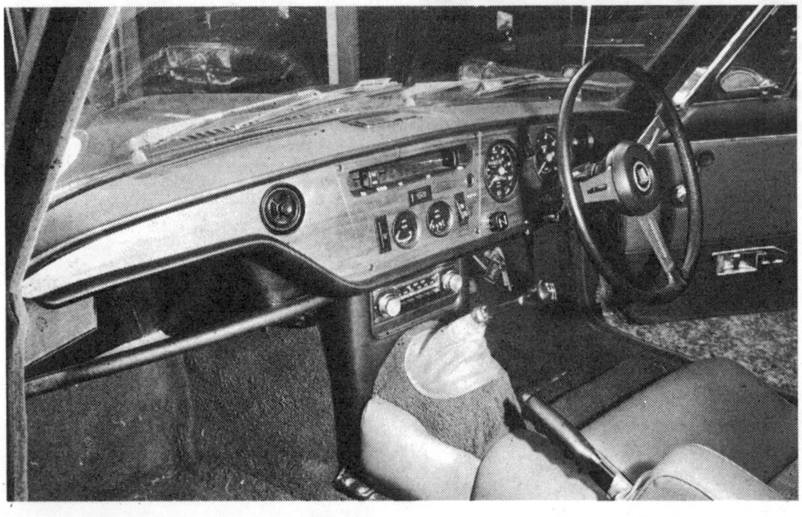

The gearbox is peculiar in that sometimes it allows very rapid, smooth changes, at other times it baulks, so you have to make two attempts to engage the desired gear. Noise was also rather apparent, our test car whining quite loudly when accelerating hard in first and second. There was also an occasional clonking from the driveshafts.

The clutch is very soft and although it allows smooth changes it was not really man enough for our standing start accelerations, when it slipped momentarily before the car got underway.

Since it received its modified suspension, the handling of the GT6 has improved a lot. Now the combination of increased traction at the rear and the weight of the six-cylinder engine gives understeer which has to be counteracted by winding on more and more lock in a rather unsporting way. Despite the modifications there is still a certain amount of camber change at the rear which generally remains unobtrusive until you are forced to lift off on a fast bend; then there can still be a mild change to oversteer.

The ultimate cornering power of the GT6 is still no higher than that of many good saloons. Considering its low centre of gravity and fully independent suspension, it rolls a fair amount and generally feels less taut and responsive when pressing hard than you'd expect.

The brakes on our test car were progressive for town use, but on the open road at higher speeds they became rather heavy and didn't inspire confidence. There was a tendency for each wheel to lock up independently and thus pull the car out of line — understandably this was even more disconcerting in the wet. The handbrake, although efficient, required considerable effort to operate, partly because it is mounted too high.

The GT6 has a firm ride that's fine for motorway cruising but which becomes rather "crashy" on poor secondary roads. This is accompanied by a fair degree of wind noise and tyre roar and even body rattles on really poor surfaces, so the smoothness and quietness of the engine is generally cancelled out by disturbances from other quarters.

The GT6 is a pure two-seater as there's only a luggage space behind the front seats. These are comfortable and offer particularly good lateral support. The tight cockpit does not allow for fully reclining seats, but it is not difficult to find a comfortable position for even the longest of journeys, although we felt it would be possible to provide more legroom than exists at present.

Although the gearlever is angled so as to be almost horizontal when in top, which means you have to get your palm right underneath it to change, it is nevertheless nicely positioned in relation to the steering wheel. We didn't like the pedals so much as there is no space to rest the clutch foot and the throttle is angled far too high, making heel and toe changes almost impossible.

The minor controls consist of an odd mixture of knobs and rockers. The circular instruments, although very functional, are partially obscured by the steering wheel. Several controls are particularly difficult to operate on the Mk 3 — the new door catches, the ventilators under the dashboard, and the combined ignition switch and steering lock that is at best awkward to operate and impossible when wearing seatbelts.

We found the heating and ventilating to be quite good and an improvement over that of the previous model. There is still a tendency for the car to mist up, making the standard heated rear window a necessity rather than a luxury. Smaller drivers found the seat belts very difficult to adjust although once done they were comfortable enough to wear. However our more long-legged members found that there was insufficient adjustment to tighten the belts fully when the seats were right back.

In its present form the GT6 offers good value for money — at £1224 it is some £190 cheaper than the MGB GT for instance. Its strong points are performance allied to economy and excellent accessibility for d-i-y-service. We feel that it is well worth paying the extra for overdrive, which allows slightly better performance in the gears as well as more relaxed cruising. ■

Motor Brief Test No 51/71 Triumph GT6

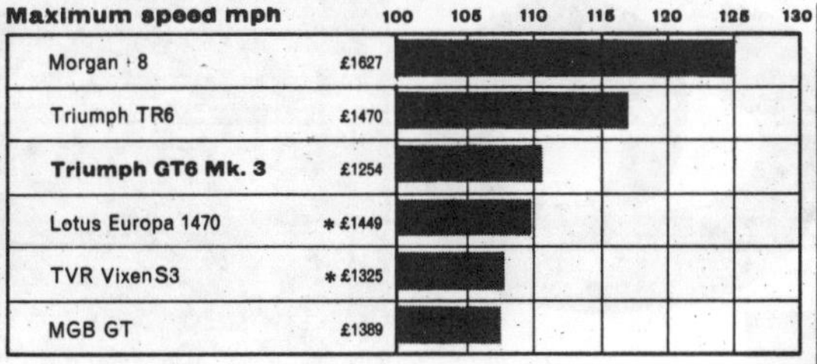

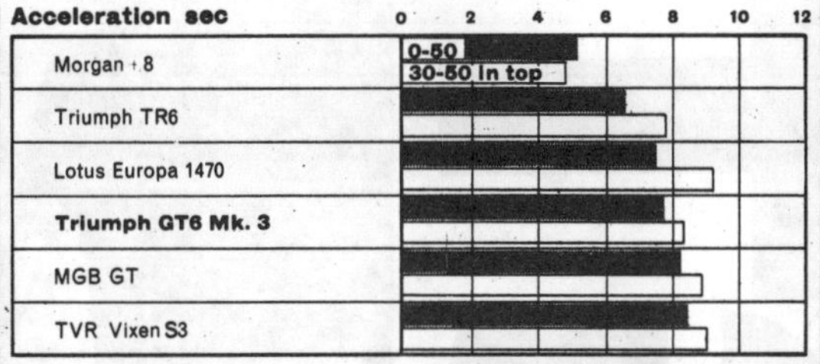

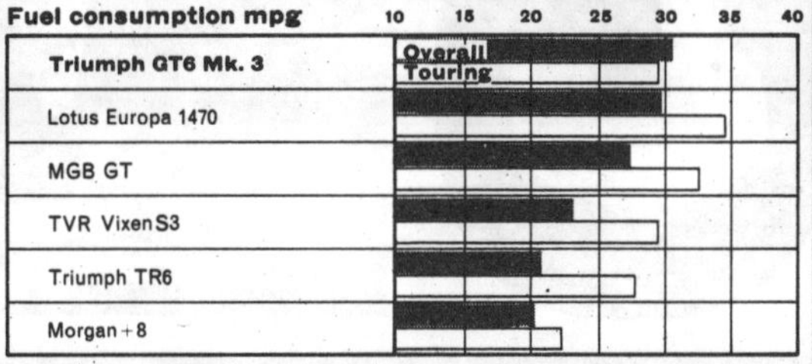

* In component form

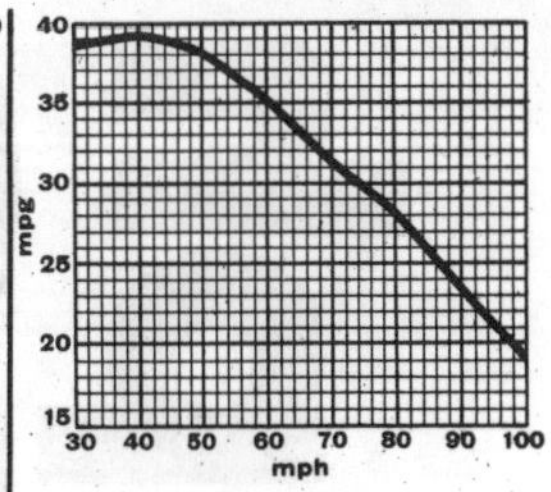

Make: Triumph. **Model:** GT6 Mk3. **Makers:** BLMC (Triumph Motor Co Ltd, Coventry). **Price:** £1002 plus £252.38 purchase tax equals £1254.38.

Performance tests carried out by *Motor's* staff at the Motor Industry Research Association proving ground, Lindley.

Test Data: World copyright reserved; no unauthorized reproduction in whole or in part.

Conditions

Weather: Dry, sunny; Wind SE 8-12 mph
Temperature: 44-60°F
Barometer: 31.5 in. Hg.
Surface: Dry tarmac
Fuel: Super premium 101 octane (RM) 5 Star rating

Maximum Speeds

	mph	kph
Mean lap banked circuit	110.5	177.7
Best one-way ¼-mile	111.1	178.6
Direct top gear		
3rd gear } at 6000 rpm	96	154
2nd gear } at 6000 rpm	68	109
1st gear } at 6000 rpm	46	74

"Maximile" speed: (Timed quarter mile after 1 mile accelerating from rest)

Mean	107.2
Best	108.4

Acceleration Times

mph	sec.
0-30	3.9
0-40	5.4
0-50	7.7
0-60	10.0
0-70	13.7
0-80	18.2
0-90	23.9
0-100	33.9
Standing quarter mile	17.6
Standing kilometre	32.6

mph	Top sec.	3rd sec.
10-30	—	8.3
20-40	8.6	7.8
30-50	8.3	6.3
40-60	8.1	6.2
50-70	8.9	6.9
60-80	10.1	8.3
70-90	11.8	—
80-100	15.8	—

Speedometer

Indicated	30	40	50	60	70
True	30	40	50	59	68.5
Indicated	80	90	100		
True	78	88	97		

Distance recorder 2.4% fast

Weight

Kerb weight (unladen with fuel for approximately 50 miles) . 17.81 cwt
Front/rear distribution . . . 56/44
Weight laden as tested . 21.51 cwt

Fuel Consumption

Touring (consumption midway between 30 m.p.h. and maximum less 5% allowance for acceleration) 29.6 mpg
Overall . . . . 30.5 mpg
(=9.26 litres/100km)
Total test distance 1,142 miles

Engine

Block material	Cast iron
Head material	Cast iron
Cylinders	6 in line
Cooling system	Water
Bore and stroke	74.7 mm. (2.94 in.) 76 mm. (2.992 in.)
Cubic capacity	1998 cc. (122 cu.in.)
Main bearings	4
Valves	Pushrod ohv
Compression ratio	9.25:1
Carburettor(s)	2 Stromberg 1.5 CD
Fuel pump	Mechanical
Oil Filter	Full flow
Max. power (net)	98 bhp at 5300 rpm
Max. torque (net)	108 lb.ft. at 3000 rpm

Transmission

Clutch	8½ in. sdp diaphragm
Internal gear box ratios	
Top gear	1.00:1
3rd gear	1.25:1
2nd gear	1.78:1
1st gear	2.65:1
Reverse	3.10:1
Synchromesh	All forward ratios
Final drive	Hypoid bevel 3.27:1

Mph at 1000 rpm in:–

top gear	20.1
third gear	16.1
second gear	11.3
first gear	7.6

Chassis and body

Construction
Separate chassis with bolted on body

Brakes

Type	Disc/drums
Dimensions	Discs 9.7 in. dia; drums 8 in. dia.

Suspension and steering

Front	Independent by wishbones and coil springs with anti-roll bar
Rear	Independent with transverse upper leaf spring, lower wishbones and radius arms
Shock absorbers:	
Front:	Telescopic
Rear:	Telescopic
Steering type	Rack and pinion
Tyres	155 SR13 Dunlop SP68
Wheels	13 steel disc
Rim size	4½J

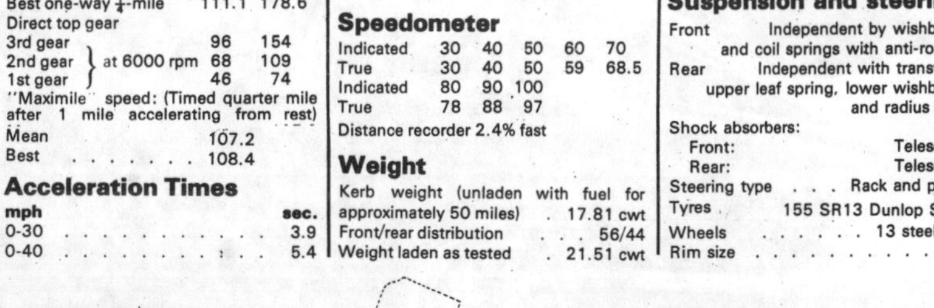

triumph GT6

Australia's cheapest two-seater coupe has developed into a remarkably refined machine but Warren Walsh has a few reservations . . .

SPORTS CARS are getting rarer these days. The market is diminishing and increasingly tough safety regulations are persuading manufacturers to forget about them.

At the same time the super-sedans and four-seater coupes are rapidly becoming the fastest, best-handling machines around.

Which makes things a bit tough for the prospective purchaser of a two-seater fixed-head coupe. In fact he has only 11 models to chose from. One costs $29,000. Three others cost well over $10,000. Another five cost in excess of $5000 which is still a lot of money for most people.

The average driver isn't keen to pay more than $2500 per seat — which gives him only two cars to chose from . . . the Triumph TR6 at $4832 and the Triumph GT6 at $4071.

The TR6 also comes in a drop-head form so it doesn't really qualify as a coupe. Which cuts the choice down to color preferences for the GT6.

When our test car popped up, we checked back to see when we last tested a GT6. To our surprise it was September 1967 — nearly five years ago.

The 67 report was on the Mark One — and the present test car is a Mark Three version so there were bound to be some improvements.

Our earlier report started tersely: "Be warned. The Triumph GT6 suffers from the sudden onset of the dreaded swing-axle oversteer. You must keep the power on through a corner or the back end will act like it is trying to nestle under an armpit."

Stirring stuff indeed. The dreaded Triumph Herald rear suspension in the Spitfire body with a 2000 engine had struck our road testers hard.

Fortunately it also struck the Triumph engineers. At the end of 1968 they cured the problem.

Thinking fast, they fitted a transverse leaf spring over the back axles. This acts like the upper arm in an A-frame (or wishbone) suspension with a new revised lower A-arm and a semi-trailing arm for better location. The driveshaft is on rubber "doughnut" couplings.

And it works — the old swing-axle lift is gone completely. No longer is the GT6 a streamlined trap for the unwary because the transition from plough understeer to power tail-out oversteer is smooth and progressive.

Even a brand new P-plater will have no trouble as long as he doesn't hit the gas pedal too hard.

)VE: Flat-out on a slow er the GT6 displays little but lots of understeer.

T: Everything is hinged he car allowing easy access most every part of the

HT: This wide-angle shot 'y shows the bulges that power.

OW: The new tail brings GT6 in line with the rest e Triumph range. All vehicles now have a Stag- treatment.

Late last year the Mark Three version of the GT6 was released in England. Externally, the seams on top of the front mudguards have gone, the rear section has been "Stagized" with an undercut tail and a new rear lamp cluster has been added.

New seats and a flat alloy leather rim steering wheel are the obvious differences inside. Somewhere in the middle of the Mark Two series, the rear luggage floor was cut away to allow for reclining seats.

This has an excellent side-benefit — the resultant wells are perfect for shopping bags. They fit exactly and stay in the one position even under the hardest lateral forces.

Face level vents and rocker switches complete the obvious Mark Three changes.

In their restyling, the engineers added another two inches to the depth of the windscreen which tall drivers will find useful.

Three safety items virtually complete the Mark Three GT6: a four-way hazard flasher, more interior padding and new steel wheels replacing the old wire ones.

Good news for owners is the steering column lock. Bad news for owners is the steering column lock position. It is under the steering column and under the dash. This makes it difficult to operate and inaccessible with seat belts on.

With most cars, the initial impressions are misleading. They wear off as the miles are piled on. But not with the GT6, our initial impressions remained right through the five days we had it.

Two things were immediately obvious; first it is a small car. It is 4ft 9in wide and only 12ft 7in long or 4ft shorter than a HQ sedan.

The inside is even smaller. Getting in isn't much of a problem, provided you follow the accepted method of left leg first into the car followed by your body and swing your right leg in last.

This method works well for both

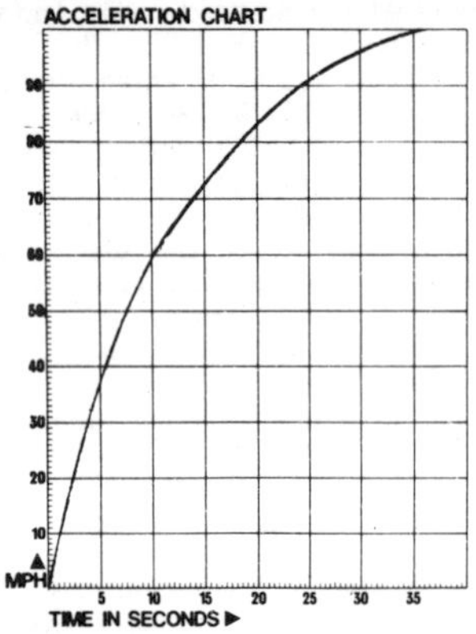

MODERN MOTOR road TEST

WARRANTY, INSURANCE, MAINTENANCE, RUNNING COSTS

Registration: **$100.50**

Insurance:
Quoted rates are for drivers over 25 with 60 percent no-claim bonus and where the car is under hire purchase. This is the minimum premium level — decreasing rates of experience and lower age groups may have varying excesses and possibly premium loadings.
Non-tariff companies $115.19
NRMA $145.05
Tariff companies $121.30

Warranty:
Six months or 6000 miles. Covers all parts and labor charges for defective materials, components or workmanship. Components from outside suppliers, such as tyres, batteries, etc. covered by their own manufacturers.

Service:
A Service .... Free
This covers the first 1000 miles (1600 km) or first month and includes lubrication and maintenance service. Materials are chargeable.

1-19 Services
There are three kinds of service:
Every 3000 miles — basically an inspection service for which the charge is approximately $6.40 labor plus parts.
Every 6000 miles — maintenance service — duration 3½ hours — $24.50 labor plus parts.
Every 12,000 miles — major maintenance service — duration 5 hours — $35.00 labor plus parts.
Oil change every 3000 miles (4800 km)
Some chassis lubrication is required.

Spare Parts — Recommended Cost Breakdown

Disc pads	$11.42
Clutch plate	$24.15
Pressure plate	$25.20
Windscreen	$25.00
Muffler	$26.45
Inner front wheel bearings	$10.00 set
Shock absorber front unit	$43.70
Shock absorber rear unit	$16.10
Tail lamp unit	$2.00
Headlamp unit	$3.20

Workshop Manuals:
Approximate cost $12.00 available through authorised Triumph dealers.

Color Range: (Upholstery range in brackets)
White (blue, tan, black) Emerald (grey, black) Sapphire Blue (grey, blue or black), Pimento (black) Saffron (black) Damson (grey, tan, black), Sienna (tan).

Minimum garage width:
Measured car width
plus one fully open door .... 7ft 2.5in (220.7 cm)

ROAD TEST DATA — SPECIFICATIONS

Importers: Australian Motor Industries, Salmon Street, Port Melbourne.
Make/Model: .... Triumph GT6
Body type: .... 2-door roadster
Pricing: as tested: $4163
basic: $4071
options/prices: .... Electric Overdrive $192
Test car supplied by: Larke Hoskins Pty Ltd, 535 Riley Street, Surry Hills.
Mileage start/finish: 2037-2572

ENGINE

Cylinders: .... Six in line
Bore x stroke: .... 2.94in x 2.99in (74.7mm x 76mm)
Capacity: .... 1998cc (122 cu in)
Compression: .... 9.25 to 1
Aspiration: .... Two sidedraft Stromberg 1.50 CD
Fuel pump: .... Mechanical
Fuel recommended: .... 100 Octane
Valve gear: .... Overhead
Max. power (gross): .... 98 bhp @ 5300 rpm
Max. torque: .... 108.3 lb.ft (16.4 kg.m) @ 3000 rpm

TRANSMISSION

Type/locations: Four speed manual floor change with overdrive
Clutch type: Hydraulically operated single plate diaphragm

Gear	Direct Ratio	Overall Ratio	MPH/1000	(KPH)
1st	2.65	10.31	6.64	10.68
2nd	1.78	6.92	9.88	15.90
3rd	1.25	4.86	14.07	22.64
4th or 3rd O'drive	1.00	3.89	17.59	35.38
4th O'drive	.80	3.11	21.99	35.38

Final drive: 3.89

CHASSIS AND BODY

Type: .... Unitary
Weight as tested incl. fuel, oil, water (no occupants): 2089lb (948 kg)
Distribution front/rear: .... 54.6/45.4 percent
Kerb weight: .... 2030lb (921 kg)

sexes. The average female won't get arrested for indecent exposure because the critical area is covered by the door.

Once seated, the medium height driver is comfortable. Short and tall people will find they are alternatively too low or too cramped. But the GT6 still fits people between 5ft 2in and 6ft 2in tall.

All the major controls are in the right places. The pedals are off-set to the left and are fairly close together. This means the driving position seems a bit awkward for the first mile or so but this effect soon wears off. Heel-and-toe downshifting is easy.

The existing 15in steering wheel is a little too large. Because the cockpit is close-fitting, big arm movements are out of the question. The present wheel sometimes requires some big movements — a smaller wheel would eliminate this problem.

On paper, the 4.2 turns lock to lock seems insane for a sports car. But the turning circle at full lock is a tiny 25ft. At full lock the front wheels are on such an angle that they scrub even at low speeds.

Steering is light at all times, and no undue reaction to potholes is felt. The general road feel is average.

The gearbox is pleasant. The movements are short and notchy and there is never any doubt about the gear positions. The syncromesh is a bit weak and fast changes are accompanied by a "snick".

Our test car came with electric overdrive on third and fourth gears. Overdrive third is nearly identical to direct fourth. This means most drivers can leave it in third gear for city work, using overdrive third as a fourth gear — without having to touch the clutch. On fast country runs, only fourth and overdrive fourth are needed.

Because engaging and disengaging the overdrive requires no clutch, driving effort is greatly reduced. A loose connection on the switch cut out the overdrive operation for a short period during the test and we missed the extra two gears.

71 ►

SUSPENSION

Front: Independent with upper and lower wishbones, coil springs
Rear: Independent, lower wishbones, trailing radius rods
Shock absorbers: Telescopic
Wheels: 4½J x 13
Tyres: Dunlop SP 68 155 SR 13
Pressures: 24lb front/28lb rear

STEERING:

Type: Rack and pinion
Turns lock to lock: 4.25
Wheel diameter: 15in (38.5 cm)
Turning circle, between kerbs: 25.2ft (7.7m)
between walls: 26.9ft (8.2m)

BRAKES:

Type: Disc front/drum rear
Dimensions: 9.9in (24.6cm), 8in (20.3cm)
Swept area: 260sq in (1687 cm²)

DIMENSIONS:

Wheelbase: 83in (211cm)
Track, front: 49in (124.5cm)
rear: 49in (124.5cm)
Overall length: 12ft 5in (388.6cm)
width: 4ft 10.5in (148.8cm)
height: 3ft 11in (119.5cm)
Ground clearance: 4in (10.2cm)
Overhang, front: 2ft 8in (81.3cm)
rear: 2ft 11in (88.9cm)

EQUIPMENT:

Battery: 12 Volt 56 A/H
Alternator: 336 watts
Headlamps: 60/45 watts
Jacking points: 4 under body

CAPACITIES:

Fuel tank: 9.75 gallons (44.3 litres)
Engine sump: 9 pints (5.1 litres)
Final drive: 1 pint (.57 litres)
Gearbox: 1.5 pints (85 litres)
Water system: 11 pints (6.2 litres)

PERFORMANCE

Test conditions for performance figures; Weather: Fine
Wind: Nil
Humidity: 54 percent
Max. Temp. 77 degrees
Surface: Dry hotmix
Top speed, average: 106 mph (169.6 kph)
best run: 107 mph (171.2 kph)
Standing Quarter Mile, average: 17.3 seconds
best run: 17.2 seconds

Speed at end of Standing Quarter: 78 mph (124.8 kph)
0-30 mph: 3.9
0-40 mph: 5.5
0-50 mph: 7.6
0-60 mph: 10.0
0-70 mph: 14.1
0-80 mph: 18.3
0-90 mph: 24.4
0-100 mph: 35.9

Speeds in gears:

Gear	Max. mph Drive	(Kph)	rpm
1st	39	64	6000
2nd	59	95	6000
3rd	84	135	6000
4th or OD 3rd	101	162	5750
4th OD	106	170	4800

Acceleration holding gears:

	2nd	3rd	4th or 3rd O/D
20-40	4.5	6.7	8.7
30-50	4.5	6.6	7.8
40-60	4.7	6.6	8.2
50-70	—	7.1	9.2
60-80	—	8.4	10.0

Fuel consumption:
Average for test: 28.4mpg (11.1 kpl)
Best recorded: 33.5mpg (11.9 kpl)
City average: 26.5mpg (9.9 kpl)
Country cruising: 30.2mpg (11.7 kpl)

Fuel flow readings (constant speeds):
30 mph: 47.6 mpg (17 kpl)
40 mph: 47.4 mpg (16.9 kpl)
50 mph: 45.8 mpg (16.3 kpl)
60 mph: 41.3 mpg (14.7 kpl)

Braking: Five crash stops from 60 mph

Stop	G	Pedal
1	.90	73 psi
2	.90	75 psi
3	.92	77 psi
4	.93	76 psi
5	.89	73 psi

30-0 mph: 1.8
60-0 mph: 3.4

Calculated Data:
Bhp/ton: 105 bhp/ton
Piston speed at max rpm: 2641 ft/min (805.2 m/min)

Speedo Corrections:

20	30	40	50	60	70	80	90
20	29	39	50	60	70	80	90

TRIUMPH SPORTS CARS

What's available to make your Spitfire, GT6 or TR go faster, handle better and look nicer

BY JOHN DINKEL
Engineering Editor

THE SUCCESS OF Triumphs in competition is due in no small measure to the amount of time, money and talent that has been invested by British Leyland in modifying its cars to go fast. Triumphs have garnered more than their share of SCCA production-class victories and national titles; the white hordes of Bob Tullius' Group 44 cars in the east and the equally overpowering Kas Kastner Triumphs on the west coast are well known by all those who are unfortunate enough to compete in the same classes.

Anyone who contemplates preparing a Triumph for competition—whether slaloms, autocross, full-bore SCCA club racing or any other form of off-highway endeavor—should consider the following advice mandatory: buy a copy of the workshop manual for your particular model. Don't just read it. Study it. Then spend $2.50 and order the appropriate book of the "Triumph competitor's Bible," the Competition Preparation Manuals written by Kastner and the present West Coast Competition Advisor, Jim Coan. There are books for every model being campaigned except the TR-2: Spitfire, TR250/TR-6, the various GT6 versions, and TR-4/TR-4A (the TR-4 manual also applies to the TR-3). And they embody all the experience and know-how that has been gathered in modifying Triumphs to run faster and handle better. You'll learn which stock engine components are strong enough for racing, which parts need reworking, what size valves to fit, how large to make the ports, what cam to use, and which sparkplugs work best. For the chassis you'll find out what tire pressures to use, what springs and shocks to fit, whether an anti-roll bar will improve handling, etc. You'll get advice on alternate gear ratios and limited-slip differentials. And just as importantly you'll find that the parts aren't available only to the chosen few: competition parts that have proved successful under the rigors of racing are available to everyone from any Triumph dealer. A list of parts for the Spitfire is found in Fig. 1. Lists for other models are available from Triumph.

Engine Testing

HUNDREDS OF hours of dynamometer testing are required to find the right combination of engine modifications and accessories that make horsepower. I spent a week with Jim Coan at the Roy Woods Racing facilities in Gardena, Calif. as he "dynoed" the new 1500 Spitfire engine to find out how much of his prior Spitfire experience could be applied to this larger engine. Starting with a stock engine (right out of Coan's own Spitfire) he first measured power as a baseline. That done, Coan then tried various combinations of carburetion, manifolding, compression, cams, calibrations and so forth—altogether 28 different tests. Fig. 2 is a chart of 16 of the runs, Fig. 3 a graph of power vs engine speed for the stock and final configuration. The results are broken up into four groups, beginning with modifications that don't require dismantling the engine and progressing onward to those changes the serious SCCA competitor would consider. Notice that though most modifications yielded more power, a few gave less. Even these "fail-

Fig. 1

SPITFIRE PERFORMANCE PARTS FROM TRIUMPH

Part No.	Description	Price
V.175	Camshaft, street-race	$ 58.80
V.110	Camshaft, type A-6	58.80
V.624	Camshaft, type A-7[1]	58.80
V.687	Camshaft, type A-8	58.80
V.549	Valve springs, retainers	32.50
V.168	Pushrods	16.00
V.438	Pistons, .040 oversize[2]	48.00
V.616	Header system[1]	87.50
307270	Header (std Mk II)[2]	50.00
V.619	Velocity stacks (2)	16.95
V.401	Gears, close-ratio	185.00
511401	Overdrive kit	395.35
502018	Ring & pinion, 4.55	46.55
502017	Ring & pinion, 4.87	42.75
508905	Nose section, 4.55	142.50
508904	Nose section, 4.87	142.50
V.391	No-spin diff, 4.10	203.50
V.174	No-spin diff, 4.55	203.50
V.393	No-spin casing, 4.10	32.00
V.392	No-spin casing, 4.55	32.00
V.525	No-spin side gears	139.00
V.451	Individual side gears	ea 12.00
144651	Inner axle, comp	50.60
V.170	Camper compensator	29.95
V.340	Front shock, Koni	26.50
V.341	Rear shock, Koni	26.50
V.399	Front springs, pair	38.50
V.400	Rear spring	75.00
512539	Competition brake pads	17.45
V.177	Oil cooler kit	69.75

[1]Mk III
[2]Mk I, II

Fig. 2

DYNAMOMETER TESTS
Engine Speed, rpm

Test	2000	2500	3000	3500	4000	4500	5000	5500	6000	6500	7000
A	26	35	43	50	57	60	61	60	59	—	—
B	26	34	43	51	57	61	64	62	59	—	—
C	25	32	40	46	56	64	68	66	64	—	—
D	27	36	44	53	60	66	69	71	72	—	—
E	26	36	44	49	54	59	58	58	56	—	—
F	27	37	46	53	57	62	64	67	60	54	—
G	28	36	45	54	62	68	72	71	69	65	—
H	27	37	44	53	61	69	75	76	74	73	—
I	26	36	45	56	64	71	77	80	80	78	—
J	27	37	39	59	68	74	79	83	84	83	—
K	29	38	50	62	70	75	82	87	89	86	—
L	31	38	50	60	68	74	78	81	83	80	—
M	31	39	52	55	71	75	80	84	84	81	—
N	30	39	48	57	65	70	74	77	77	72	—
O	29	39	49	58	64	67	69	69	68	66	—
P	—	37	47	59	68	75	81	87	92	95	90

Triumph's West Coast Competition Advisor, Jim Coan, ran the dyno and turned the wrenches during testing of the 1500 Spitfire engine.

Fig. 3

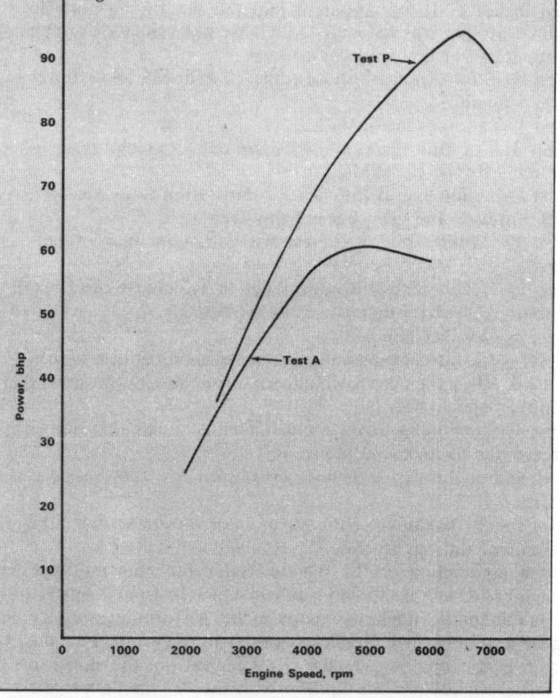

Fig. 4

ACCELERATION, STANDING START

	Stock Engine	Modified Engine	Modified Engine plus Electric Fuel Pump
0-60 mph, sec	15.4	10.9	9.7
0-80 mph, sec	31.8	29.5	18.8
0-¼ mi, sec	20.2	17.7	17.4
Speed at end of ¼ mi, mph	68	72	78

TRIUMPH

ures" are information successes, however, as they tell a lot about how an engine responds to various tuning techniques. Although the numbers shown apply only to the Spitfire 1500, about the same percentage increase in power can be expected from other Triumph engines with similar modifications. Here are the changes that were made for each test:

Group 1: Initial testing was done on the standard engine with stock compression ratio, ignition timing, 1.5-in. Stromberg carburetor, cast iron exhaust manifold and muffler.

Test A—Various ignition settings were tried with no success over the stock setting.

Test B—Test without muffler; this test indicates the stock muffler is very effective with the standard engine configuration.

Test C—Stromberg carburetor and manifold replaced with twin 1¼-in. SU carburetors and manifold from Spitfire Mk III.

Test D—Triumph V.616 tube headers installed.

Test E—Headers with Stromberg carburetor.

In all these tests valve float developed at 5800-6000 rpm.

Group 2: For this series of tests the head was milled 0.150 in. for a compression ratio of 9.7:1. Stock valve springs were replaced with the competition ones, part number V.549.

Test F—Engine as noted just above. Engine now revved easily to 6500 rpm, thanks to the valve springs.

Test G—Intake manifold changed to Mk IV manifold with larger inlet ports. Note power increase above 3500 rpm compared to test F. It was apparent that the size of the ports in the 1500 Spitfire was reduced to increase gas velocity through the manifold for emission control.

Test H—Stromberg carburetor replaced with Mk III carburetors and manifold.

Test I—Tube headers installed.

Group 3: For this series of tests the cam was changed to Triumph's B-type camshaft.

Test J—Engine with B cam, 9.7:1 compression ratio, headers, SU carburetors and glass-packed muffler.

Test K—Muffler removed; test with straight-pipe configuration.

Test L—Head milled additional 0.060 in. to achieve compression ratio of 10.5:1, otherwise same as Test J.

Test M—Muffler removed.

Test N—Cast iron exhaust manifold, 1½-in. Stromberg carburetor, Mk III intake manifold (interchangeable with Stromberg manifold) and muffler.

Test O—Intake manifold from 1500 installed. Test shows decrease due to manifold alone.

Group 4: For this series of tests Triumph's A-6 camshaft was installed.

Test P—This is the ultimate power combination tested: tube headers and the Spitfire Mk III's induction system.

Upon completion of the dynamometer runs the modified engine (Test P modifications) was reinstalled in Coan's Spitfire and taken to the track to compare the performance of the original engine (which R&T had tested in May '73) with the "race" version. Fig. 4 gives the numbers but not the story: in the initial shakedown the Spitfire with the "P" mods was 4.5 sec quicker to 60 mph and ran through the ¼ mile 4 mph faster and 2.5 sec quicker than the stocker. But performance above 65 mph was disappointing: as the graph shows the hot Spitfire didn't reach 80 mph much sooner than the stock engine, and above 80 mph the stock Spitfire would run away from the tweaked car. Something was definitely amiss, so we tried an acceleration run from a steady 70 mph to 80 mph with time allowed for the fuel bowls to fill before opening up the throttle. This resulted in 70-80 mph times chopped in half, so it was evident the engine was starving for fuel at high speed during a wide-open-throttle acceleration run. Back to the drawing board for an electric pump!

Fig. 5

LATERAL ACCELERATION
100-ft radius

	Tires	Pressure f/r, psi	Lateral Accel, g
Spitfire 1500	5.20-13 bias	21/26	0.720
	165-13 radial	28/32	0.758
GT6	155-13 radial	24/30	0.704
	165-13 radial	34/32	0.745

Two days later Coan was back and the results were more like we expected. Quarter-mile times were cut an additional 0.3 sec and the speed increased 6 mph. The engine was now making horsepower at top end. Acceleration time to 60 mph was chopped 1.2 sec but even more dramatic was the reduction in 0-80 mph time: slashed more than 10 sec from the same engine without the electric pump. And the modified engine was still pulling strongly long after the stock one had peaked.

Just to set the record straight, however, we must remind you that with such modifications emission control goes out the window, so although the extra urge may be immensely satisfying to the driver its results are not kind to the environment and are not legal. For racing use they are at least legal.

Spitfire Handling

For owners of early Spitfires (Mk I, II and III) the best investment for improved handling is the installation of the rear "swing" spring from Mk IV and 1500 models. The Spitfire's arrangement of swing axles and transverse leaf spring rear suspension derived from the Herald sedan, like most old-style swing-axle setups with their extremely high roll center, likes to jack up the rear of the car, get the wheels all cambered and send the rear end sliding. The way to minimize this is to reduce the roll stiffness of the suspension, and that's exactly what camber compensators do.

Triumph engineers, however, found a way to modify the Spitfire's transverse spring that would cut its roll resistance by three-quarters. Only the main spring leaf is clamped to the differential so that it can contribute roll stiffness; the other leaves are free to rock at their centers so they contribute practically nothing but bump stiffness. Furthermore, the whole spring assembly has a higher 2-wheel bump rate so that camber change under braking and acceleration is reduced. To make up for the lost roll stiffness at the rear, the front anti-roll bar was enlarged from 0.69 in. to 0.88. Both the rear spring and the larger front bar bolt right into earlier models and the improvement in cornering power and transient response is considerable. Early Spitfires couldn't outhandle a good small sedan; now they're definitely in the sports car class.

Tire Testing

One of the questions we are asked most frequently is how much improvement in cornering can be gained from better tires and higher tire pressures. We just happen to have run some skidpad tests on the Spitfire and GT6 (Fig. 5) because both cars were part of our April 1973 9-car Showroom Stock Sports Car tests. Rim size was the same for each car regardless of the tire fitted so the tire size, pressure and type (in the case of the Spitfire, we tried bias and radials) were the only variables. Semperit radials had proved quite "sticky" when tested against eight other radials last August, and the improvement they or other premium tires can give is considerable in both cases.

Similar results should be obtained not only with Triumphs but other cars as well. Don't assume, however, that because big is good, bigger is better. One or a maximum of two sizes larger than standard is the biggest increase we'd recommend for any car other than a full competition model, on which other changes are made to accommodate them or get the most from them.

Aftermarket Accessories

THE LIST of competition parts sold by Triumph dealers is weighted heavily toward SCCA competition. Obviously it would be prohibitive for Triumph to test, stock and sell parts for every type of competitive event that Triumphs can enter, as the rules vary so widely among various sanctioning organizations. That's where the aftermarket manufacturers come in. Need a Weber carburetor and manifold for circle-track racing? A cam for drag racing? Or special springs for bounding around Baja? Aftermarket suppliers listed in Fig. 6 can supply almost all your needs for these hard-core racing parts. In addition you can find all sorts of add-on accessories to make road driving safer and more enjoyable—from rollbars to anti-roll bars, steering wheels to road wheels, bucket seats to front spoilers. Get yourself a few catalogs, decide where you want to compete and how fast you want to go, and pick from the wide variety of equipment available from specialty manufacturers and Triumph dealers. You should be well on your way to a fast competitive Triumph or a more enjoyable Triumph for the road.

Fig. 6

SOURCES OF MANUALS & EQUIPMENT

GENERAL

Workshop and Repair Manuals
Various Guides available from:
Autobooks, 2900 R W. Magnolia, Burbank, Calif. 91503
Carbooks, 2628 Atlantic Ave, Brooklyn, N.Y. 11207
Classic Motorbooks, 3844 Thomas Ave South, Minneapolis, Minn. 55410
Robert Bentley, Inc, 872 Massachusetts Ave, Cambridge, Mass. 02139

Tuning Manuals
Triumph Competition Preparation Manuals ($2.50)
East of Miss.: Mike Barratt, British Leyland Competition Dept, 600 Willow Tree Rd, Leonia, N.J. 07605
West of Miss.: Jim Coan, British Leyland Competition Dept, PO Box 1557, Gardena, Calif. 90249
(Both of the above sources will supply additional technical information and advice on Triumph preparation.)
Triumph Sports Owners Association Bulletin: TSOA, 600 Willow Tree Rd, Leonia, N.J. 07650
(A monthly newsletter that includes technical tips, membership $5.00)

Performance Accessories and Racing Equipment
The following companies sell all types of engine, suspension and misc. performance equipment:
Autoworld, Dept 23, 701 N. Keyser Ave, Scranton, Pa. 18508
B&B Auto Sport Ltd, 150 C Lakehill Rd, Burnt Hills, N.Y. 12027
Cannon Industries, Inc, 9067 Washington Blvd, Culver City, Calif. 90230
D&E Competition, 6215 Lancaster Ave, Philadelphia, Pa. 19151
Honest Charlie Inc, Box FC 8535, Chattanooga, Tenn. 37411
James Auto Specialties, 3820 W. Magnolia Blvd, Burbank, Calif. 91505
J.C. Whitney & Co, 1917-19 Archer Ave, Chicago, Ill. 60616
King Motoring Specialties, 6704 Crescent Blvd, Pennsauken, N.J. 08110
MG Mitten, 36 S. Chester, Pasadena, Calif. 91106
Moon Equipment Co, 10820 S. Norwalk Blvd, Santa Fe Springs, Calif. 90670
Motor Racing Equipment, Inc, 1412 Borchard Ave, Santa Ana, Calif. 92705
Pacer Performance Products, 5345 San Fernando Rd W., Los Angeles, Calif. 90039
Sears, Roebuck & Co.
Vilem B Haan, Inc, 10305 Santa Monica Blvd, W. Los Angeles, Calif. 90025
Wilco, PO Box 1128, Rochester, N.Y. 14603
Wrep Industries Ltd, 2965 Landwehr Rd, Northbrook, Ill. 60062

ENGINE

Complete Engines, Blueprinting, Dyno Tuning
PAECO, 213 S 21st St, Birmingham, Ala. 35233
(Engines available in kit or assembled form in 4 stages of tune)

Camshafts
Crane Cams, 100 NW 9th Terrace, PO Box 160-50, Hallandale, Fla. 33009
Crower Cams & Equipment Co, 3333 Main St, Chula Vista, Calif. 92011
Sig Erson Racing Cams, 20925 Brant Ave, Long Beach, Calif. 90810
Iskenderian Racing Cams, 16020 S. Broadway, Gardena, Calif. 90247
Norris Performance Products, 14754 Calvert St, Van Nuys, Calif. 91401
Racer Brown Inc, 9270 Borden, Sun Valley, Calif. 91352
Weber Speed Equipment, 310 S. Center St, Santa Ana, Calif. 92703

Carburetors and Manifolds
Dellorto sidedraft carbs: Tabline, 1-A Orchard Ln, Berkeley, Calif. 94704
Interpart, 100 Oregon St, El Segundo, Calif. 94245
SU carb tool kits: available at imported-car parts stores and specialty shops
Uni-Syn: available at imported-car parts stores and specialty shops
Weber carbs and manifolds: Jack McAfee Imports Inc, PO Box 1056, Burbank, Calif. 91505
British Tuning Center, 486 Thompson Ave, Mt. View, Calif. 94040

Cylinder Heads—Porting, Polishing, Blueprinting
Mondello Industries Inc, 1666 Euclid, Santa Monica, Calif. 90404
Valley Head Service, 18422 Oxnard, Tarzana, Calif. 91356

Exhaust Systems
Abarth exhaust systems: available at imported-car parts stores and specialty shops
Stebro exhaust systems: available at imported-car parts stores and specialty shops

Headers
Clifford Research Co, 774 Newton Wy, Costa Mesa, Calif. 92627
Exzostec Inc, 16100 Gundry Ave, Paramount, Calif. 90723
Hooker Headers, 1032 W Brooks St, Ontario, Calif. 91762

Pistons
Forgedtrue Corp, 1480 Adelia Ave, S. El Monte, Calif. 91733
Jahns Quality Pistons Inc, 2662 Lacy St, Los Angeles, Calif. 90031
J. E. Engineering Corp, 930 Monterey Pass, Monterey Park, Calif. 91754
Venolia Piston & Ring Co, 2160 Cherry Industrial Circle, Long Beach, Calif. 90805

BRAKES

Brake Pads
Ferodo pads: Elsco Corp, 1645 Jessie St, Jacksonville, Fla. 32206
Girling pads: Joseph Lucas N. America Inc, Dept RT, 30 Van Nostrand Ave, Englewood, N.J. 07631
Lakewood pads: Ladewood Industries, 4800 Briar Rd, Cleveland, Ohio 44135
Repco pads: Repcoparts USA Inc, 6281 Chalet Dr, Los Angeles, Calif. 90040

SUSPENSION

Anti-roll Bars
Addco, 60 Watertower Rd, Lake Park, Fla. 33403
Scona, 807 N Vermilion, Danville, Ill. 61832

Shock Absorbers
Koni shocks: available from imported-car parts stores and specialty shops
Spax shocks: available from imported-car parts stores and specialty shops.

Wheels
Ansen Automotive Products, 13715 S. Western Ave, Gardena, Calif. 90249
Appliance Industries, 23902 S. Vermont Ave, Harbor City, Calif. 90710
American Racing Equipment, 540 Hawaii, Torrance, Calif. 90503
Cragar Industries, 19007 S Reyes Ave, Compton, Calif. 90221
T. Hoff Inc, 1109 N West St, Raleigh, N.C. 27603
Minilite mags: Hank Thorp Inc, PO Box 201, Edison, N.J. 08817
Rocket Wheel Industries, 3501 Union Pacific Ave, Los Angeles, Calif. 90023
Western Ohio Wheels Inc, 2326 E River Rd, Dayton, Ohio 45439

ACCESSORIES

Rollbars, Front Spoilers, Steering Wheels
Available from various imported-car parts stores and specialty shops

Seats
Recaro: Nortac World Trading Co, 800 E Holt Ave, Pomona, Calif. 91767

Instruments
Stewart Warner Instrument Division, 1840 Diversey Pkwy, Chicago, Ill. 60614
VDO Instruments Automotive Division, 116 Victor, Dept R-6, Detroit, Mich. 48203

Miscellaneous
Aeroquip braided stainless steel brake lines, hoses, etc., plus aircraft surplus: Henry's Engineering Co, PO Box 2550, Landover Hills, Md. 20784

Triumph to Turin

Paul Davies took a trip in a Triumph to Turin's Motor Show — and tried a few Italian traffic light Grands Prix in a Fiat 126 at the same time.

'YOU CAME IN A GT6 — BUT IT'S 700 miles!' The old lags of motoring leaning against the bar of Turin's Majestic Hotel nearly choked on their Campari Sodas as the real truth behind the innocent statement set in. Who was this mad young fool who had braved the cobbles and autoroutes of France — not to mention the twists and turns of the Mont Cenis pass — with only a Triumph GT6 to aid him on his way?

Anyone would think it was no more than a motorised bath chair the way people carried on about my transport to the last of the annual Turin motor shows. After all, Italy isn't that far away — and really the GT6 is a quite economical and civilised means of getting from one place to another in a sporting manner. The thing is, it looks like a sort of big-engined hard top Spitfire and people seem to treat it as such. Even the men of the motoring press who should have known better. At a list price of £1373 the Triumph is not all that expensive for a 110 mph car that will climb to sixty em pee haitch in just over ten seconds. It's small, yes, but it turned out to be quite a comfortable means of transport that takes two people and an unbeliveably large amount of clobber. But more on GT6's later. First, Turin.

The Salone Internazionale dell 'Automobile Torino is the place where for many years the gapers have gone to gape at the latest from the Italian stylists — masters at the art of professional customising, so to speak — who all try and out-physch each other at the Nth hour to bring off the coup d'etat of the show. Go there on the evening before the show opens to the public and half of the styling stands (in their own hall away from the more mundane bread and butter exhibits) will be empty. Arrive next morning and the place is full with gleaming creations for future production by one of the larger volume producers, one offs, and even plaster mock ups which will probably never see the light of day again. It's all part of the Turin Motor Show, which after '72 follows Germany's national auto show and goes on a once every two years schedule. (How long before London follows?).

Apart from being the last until November 1974 the 1972 show was, I am told, no different from others. But it was the first Yours Truly had visited. The first thing that strikes one about Turin itself is that at motor show time, and most of the rest of the year in fact, everyone goes car crazy. Where else in the world would you get your car broken into and only the spares kit, a stop watch and a motoring anorak pinched? Hard luck whoever had the gear — it was a Custom Car anorak (no-one wants those) and they don't sell GT6's in Italy anyway.

The show accepted there was something special for even the Italians to get excited about. It was the launch of the Fiat 126. Now the 126, you will have heard by now, is the eventual replacement for the famous 500 baby car. And if you've ever been in Italy you'll know that everyone and his son owns a 500.

The 126 follows the general body style set a short while back by the slightly larger Fiat 127. The big difference lies in the fact that the 126 carries its power unit — a 23 bhp, 594 cc, air cooled twin, which is in fact an uprated version of the old 500 motor — at the rear. The two door body is an ample four seater and basic styling and interior trim is best described as 'adequate'. The car is larger than the 500 (which at present continues in production with the same power unit as the 126) and more modern in concept. Front suspension is independent with a transverse leaf spring and the independent rear uses coil springs and semi trailing arms. Brakes are drum all round and the four forward speed gearbox uses synchro on the top three ratios — unlike the 500 which had an all crash box. Radial ply tyres are standard equipment on the 12in diameter wheels.

A lot of safety engineering has gone into the 126. The body is extra strong, the steering column collapsible and the fuel tank safely sited under the rear seat rather than in the rather crash prone nose position of the 500.

The car is fun to drive and ideal for a crowded place like Turin. It doesn't matter that the car is flat out at around 65-70 mph or that acceleration times are best taken with a calendar, in heavy town traffic it is plenty quick and very, very, responsive. After a time I worked on the principle that if there was a gap in the traffic, it was wide enough for a 126. To someone not brung up on the 500 the 126 was very much a novelty, but to the Italians it was a highly practical form of transport and with the better looks, slightly more room and performance over the 500 could well make its mark in many other places. With the fact that a car like the 126 exists it's hard to think why others keep mumbling on about producing 'city' cars.

Big problem with the 126 as far as this country is concerned could well be price. In Italy the car will sell at over £600.00 which means that it could be nearly £800.00 in Britain by the time supplies start arriving in the middle of this year. We can but wait patiently and see.

Turin show itself was fun, but not outstanding. Vote for best looking car at the event must go to that established master Bertone who showed his Maserati Khamsin which is scheduled to go into full production during this year. The large GT car uses a five litre V8 motor and, unlike previous front engined Maseratis, uses an independent rear suspension. Most striking part of the styling of the car is at the rear end where the complete rear panel is finished in glass with the rear lights set in. This allows really good rear vision and is an idea that was previously tried on the Jaguar Pirana built specially for the Daily Telegraph mag some years back and shown at Earls Court. One interesting thing about the Khamsin was that Maserati appeared to have used their link with Citroen to utilise the power steering system first fitted to the French SM model which uses a Maserati V6 motor.

The other big item on the Bertone stand was the Lancia Stratos which is now going into production with a two litre, four cylinder, dual OHC motor instead of the Dino V6 used in the HF competition version. All of which gets very confusing when you remember the Dino unit is a Ferrari engine and the double knocker four for the street version of the Stratos will be a development of the Fiat 132 motor which is also used in the new Lancia Beta in a transverse front wheel drive layout. Fiat, of course, own Ferrari and Lancia — as well as Abarth and Autobianchi and Weber, and . . . .

Other famous name is Pininfarina who showed a hardtop version of the Alfa Romeo Alfetta with mid engine and detachable roof panel. Quite nice but finished in a really ghastly matt lime green which did nothing for the looks of the car. Another car in this strange finish (pink this time) was the FL-1 mid-engined creation from Lombardi which is made to accept either the two litre Lancia (Stratos) engine or a Ford V6 of all things. Man to make the big time in the past few years has been De Tomaso who showed various wild looking street Panteras and a new coupé called the Longchamps. If you remember the Deauville saloon looked a bit like an XJ6 you do not have to look any further than a Mercedes 350SLC to see the basic lines of the latest De Tomaso.

Put into the 'why did they bother' category must be the Varedo which was a mid

GT6 interior — well laid out, good instruments, but no oil pressure gauge. The Triumph 1998cc six-pot mill is torquey and has plenty of power.

engine design from Iso Rivolta. Using a Ford 5.7 litre V8 motor this used a glassfibre body and looked rather out of place amongst the purposeful looking Lele, Grifo and Fidia cars which are made by the same company. Iso say the Varedo will go into production but more knowledgeable types than me (at the Turin show that is) say the style has been shown several times before.

Surprise exhibit on the Ital Design stand was the Esprit which uses Lotus mechanicals. Colin Chapman has been quoted as saying production might be possible within a few years. We can but hope. Ital is the business of ace stylist Girogio Giurgiaro who used to work for Bertone and has recently done the work on the small (relative) mid-engined Maserati Merak which is a very smooth looking car.

Going away from the big names, the Giannini concern who are known for selling Fiat tuning gear, had a two litre Fiat 132 package and also a 1600 cc 128 saloon. Giannini also showed a buggy powered by a Fiat 500 engine — which seemed very tiny until the Ital Jet stand was found where the company had several 50cc buggies on show. In fact the beach buggy scene appears to be much more alive in Italy than it is over here. There were nine different makes on show, including one VW with tracks on the rear, a Skoda powered device, several different Fiat powered models and — believe it or not — one

Fiat's tiny 126 is a joy in traffic, and continues the marque's line of utility transport

called the (wait for it) 'Hot Car'. Yes really, it's made in Milan.

Of all the models on show at Turin the personal favourite — although only a plaster mock up — was the Coggiola design for a full four-seater Alpine Renault. It looked practical and beautiful but may well go down amongst those ideas that will never see a production line.

Talking of production lines, of course, brings us back to the GT6. There's little need to run through the basic specification again — since the car's introduction the only changes of any consequence have been a lower link to the rear independent suspension (to eliminate the dreaded swing axle tuck under) and a modification to the cylinder head. Stylewise the tail section now has the chopped off rear that also appears on Triumph's Stag and Spitfire, and the front wings have lost their strange vertical seams.

In latest form the car looks good. The two doors are plenty big enough and the rear hinging window (heated as standard) allows really good access to the flat luggage platform behind the two seats. The seats, which are adjustable to rake, are fairly small but hold you well and the whole driving position feels very much like a sports car — low, lie back with arms stretched and all the instruments and controls where they should be. Virtually everything that is required is there (no oil pressure gauge however) and most of it comes as standard equipment for your cash. The only real option is an overdrive which operates on 3rd and top gears and costs an extra £72.

The six cylinder, 1998 cc, motor starts quite easily on choke and runs very smoothly. It's a torquey engine that you don't need to extend to the 6000 rpm red line to get maximum performance. The bottom two gears are very low, which is ideal for quick getaways and mountain roads but means you have to watch the tacho very closely if you are accelerating quickly. The gearchange itself on our test car was very stiff and cog swapping did not seem as smooth as it should have been, the overdrive switch in the top of the knob often getting inadvertently operated during a heavy handed gear shift.

Much of the trip down through France to Turin was by motorway. Only an hour after leaving the Townsend-Thoreson ferry (speed and efficiency once again) we were on Autoroute which is where the car stayed for over 400 miles. Although the French have a 110 kph (72 mph) limit on their main roads there is, sensibly, no restriction on the motorways which means your cruising speed can be whatever you care to make it. With the GT6 in O/D top the ideal figure was just on the 100 mph. At this speed the car was well under its maximum rev limit and engine and noise level, although by no means silent, was quite bearable. The ideal speed for tramping on all day — except the overdrive solenoid started to play up and the car would suddenly jump out of O/D without any warning. This meant restricting cruising to about 90 mph so that any jump into direct drive would not over-rev the engine. Ninety is in fact about all the GT6 will comfortably sustain in plain 4th gear — over that the engine gets very fussy and extremely noisy. All of which helps to prove that if you are buying a GT6 the optional O/D should be considered an essential.

Once off the autoroute and into the mountains on the French-Italian border, the car again proved an ideal choice for the journey. Steering is light and very accurate and the small size of the car means that there's no problem at all negotiating successions of hairpins. During this trip the brakes stayed brakes — no fade, even on the long down mountain parts — and the cooling system stayed cool.

Despite what everybody seemed to think the GT6 was an ideal car for the journey. It was economical (24.9 mpg for the 1600 miles covered) fairly fast and reasonably comfortable. The only mechanical defect, apart from the jerky overdrive control, was a loose bolt on the exhaust which was most speedily cured by about fifteen Leyland-Innocenti mechanics who happed to be hanging around the Turin show. They laughed when I sat down to drive — but I rate the GT6 as the most under-rated sports car on the British market. **PD**

TRIUMPH SPITFIRE AND GT6, 1962-1970

USED CAR R&T CLASSIC

Incomparable cars in their price classes

BY THOS L. BRYANT

PHOTOS BY GORDON CHITTENDEN

LEYLAND MOTORS TOOK over the Standard-Triumph company in late 1960 and saved it from financial death. While S-T was doing fairly well with overseas sales of the Triumph TR3A sports car, the British home market was seriously depressed by a credit squeeze, with the result that fewer than 650 TR3As were sold in the U.K. in 1960 and, amazingly, less than 100 in 1961. One of the early moves following the Leyland takeover was the revival of a small sports car project which the S-T management had commissioned Giovanni Michelotti to design. It was to be competitive with the Austin-Healey Sprite and soon-to-come MG Midget; a sports car that would be less expensive than the TR3A and the upcoming TR4. Michelotti came up with what would eventually become the Spitfire, but his prototype arrived at the S-T Coventry facility when the company's finances were near rock bottom, so the prototype was pushed into a corner and tucked away. One of the early decisions of the new Leyland management was to revive the program in short order, and in late 1962 the Triumph Spitfire 4 made its debut.

One of the given design parameters for the Spitfire was that it be based on the Triumph Herald chassis and running gear, but Triumph aficionados are quick to point out that the sports car is *not* simply a Herald with a roadster body. In his book *The Story of Triumph Sports Cars*, Graham Robson writes, "Herald front and rear suspensions were also used, with minor changes to rates and damper settings, together with the rack-and-pinion steering along with a collapsible steering column which was a safety feature years before legislation was to demand it. The chassis, though similar to the Herald in concept, was very different in detail. In principle it was a backbone design, as was that of the

Herald, but in conjunction with changes to body construction the outriggers and bracing members had been deleted. The new chassis was designed around a wheelbase 8.5 in. shorter than that of the Herald, and the only common parts were suspension mountings and supports."

The Spitfire has a double Y-type frame with the front Y accommodating the engine and the rear Y tying in with the frame-mounted differential and independent rear suspension. The irs feature of the Spitfire was a major selling point because it was the only car in its price class that didn't use a solid rear axle. The Spitfire used swing axles and a single transverse leaf spring, as we pointed out in our original Technical Analysis of the car (R&T, April 1963). We went on to say, "In order to circumvent certain well known disadvantages of what is, otherwise, a very simple design, Triumph engineers have kept the roll center as low as possible (13.8 in.), used an initial wheel setting of 2° negative camber and added a trailing link to give a slight roll understeer effect. All these features, plus a very low rear-end roll stiffness (about 90 ft-lb/degree), tend to alleviate, if not eliminate, oversteer, jacking and wheel hop. A smaller [than the front] total wheel travel (5.5 in.) also helps to keep the extreme changes in camber down to reasonable proportions, the figures being 8° negative at full bump and 10° positive at full rebound."

Early Triumph Spitfire, such as the 1963 model at left, had relatively spartan interior but roll-up windows were included. Instruments were clustered in center of dash. GT6 coupe was introduced in 1966.

It all sounded very good on paper and at touring speeds the irs did give the Spitfire better-than-average ride characteristics. However, at a more spirited pace, the swing axles lived up to their name and swung, producing more wheel hop and jacking than most road testers cared for, as we pointed out in our first Spitfire test (also April 1963):

"When pressed along at racing speeds, however, the picture changes a bit. The high rear roll center causes the back of the car to lift and the wheels begin to pull under, which sends the car skating sharply outward. The Spitfire's quick, precise steering makes it easy to catch the car before the situation gets out of hand, but it is impossible to get around a race course very rapidly in a series of swings and slides. Also, on tight corners the inside rear wheel lifts completely clear of the road and spins free, making acceleration out of the turn very leisurely indeed."

It wasn't long before aftermarket camber compensators were being sold to Spitfire owners who wanted to improve the high-speed handling characteristics. Replacing the original equipment 3.5-in.-wide wheels and the 5.20-13 tires with wider ones (preferably radial tires) also made a big difference and brought more stability to the rear end.

The Spitfire's front suspension was adapted from earlier Triumph cars and had been well proven through usage in a variety of formula racing cars. It's a fairly simple design with A-arms, coil springs, tube shock absorbers and an anti-roll bar, all of which work well together and give a generous 6.5 in. of total wheel travel from full bump to full rebound.

Engine & Gearbox

THE BEST adjective for describing the Spitfire's 4-cylinder engine is rugged. It dates back to the Standard Eight sedan of 1954, but there have been a number of changes over the years. In its initial stage, the engine was 803 cc, but grew to 948 cc in the Herald and later to 1147 cc in that model and in the Spitfire when it was introduced. In the Herald or 1200, the engine generated 43 bhp (gross) at 4500 rpm, but the Spitfire developed 63 (gross) at 5800 through altered valve timing and carburetion (twin SUs).

The Mark 2 Spitfire (1965–1967) engine received a slight horsepower boost (to 67 gross) through a revised camshaft profile and a 4-branch exhaust manifold. The next update came in 1967 with the introduction of the Mark 3. Engine displacement increased to 1296 cc and the Spitfire engine was now based on the 1300 Triumph instead of the 1147-cc unit used before. The gross horsepower rose to 75 at 6000 rpm, but the following year, U.S.-version Spitfires suffered their first encounter with emission controls and the bhp figure sagged to 68 at 5500 rpm, where it remained through the term covered by this report.

The gearbox is a relatively simple affair with synchro on 2nd, 3rd and 4th gears. An electrically operated overdrive on 3rd and 4th gears was offered as an option in late 1963 and while it did not improve performance, it did improve the already impressive fuel economy under certain conditions. The Spitfire went into production with a single dry-plate Borg & Beck clutch, but this was replaced with a larger, diaphragm-type unit on the Mark 2s.

General Comments

PUTTING THE Spitfire in perspective, from its introduction in late 1962 it was a welcome addition to the small sports car range, with such features as the independent rear suspension, front disc brakes, roll-up windows, pleasing Michelotti styling and an affordable price ($2199 list in 1963). The Mark 2, introduced in 1965, was little changed except for a new extruded-aluminum grille, a "Mark 2" emblem on the rear deck lid and vastly improved upholstery material and seat padding. The Mark 2 price rose only slightly, to $2249 list, and continued to sell at a good pace (Mark 1 sales from 1962 to 1965 were almost 46,000 and more than 37,000 Mark 2s were sold from 1965 to 1967).

The Mark 3 (1967–1970) represented more sweeping changes in the evolution of the model. The already-mentioned larger engine was complemented by better braking through the use of slightly larger calipers on the front discs, and the build-it-yourself convertible top with separate frame was replaced with a lift-over-and-clamp top that was infinitely easier and more convenient to use. The most notable exterior change was a new front bumper that was characterized as giving the Spitfire the appearance of a shark with a bone in its teeth because it split the grille opening horizontally. The Mark 3's list price took another small jump, to $2373, but it was still a good buy in terms of value received and in comparison with competing sports cars such as the MG Midget III with its $2255 price tag.

Swing-axle rear suspension gave comfortable ride characteristics during normal driving but had its drawbacks in hard-driving situations. Jacking and wheel hop were common complaints of those who used their Spitfires for gymkhanas and slaloms as well as racing.

Triumph GT6 and GT6+

IT DIDN'T take long for some enthusiasts to start thinking and talking about a 6-cyl Spitfire and, in fact, conjecture about such a car began as soon as the Spitfire 4 was introduced. The decision makers at Triumph were not terribly keen on the idea because their production facilities were stretched to the limit. But it was bound to happen. Michelotti was given free rein to design a fastback body for the Spitfire and, according to Robson's Triumph book, work on the design was completed during the winter of 1963–1964. Robson writes, "It soon became clear that the added weight would not make it a very attractive proposition on the Spitfire as it stood, and for a time it was used as 'executive transport' for senior engineers."

What was to become the GT6 was revived in 1965 and the first production models appeared in 1966. The engine was the overhead-valve 1998-cc unit from the Triumph 2000 sedan, developing 95 bhp (gross) at 5000 rpm. It was mated to an all-synchro gearbox which had been developed for the Triumph Vitesse. The new, stronger transmission was deemed necessary to handle the additional torque load of the 6-cyl engine, as was a stronger rear-end assembly to handle the 3.27:1 final-drive ratio in place of the 3.89:1 Spitfire ratio. So, a new case and stronger half shafts were fitted to the GT6.

The suspension for the new car was little different from the Spitfire and not everyone was impressed with swing-axle rear suspension on a car with the torque of the GT6. The British motoring press by and large found the suspension arrangement unsatisfactory, complaining that it was too soft and that it produced unacceptable handling characteristics. Triumph's response was that the suspension's softness was designed for American customers, pointing out that more than 90 percent of the company's sport cars had been exported to the U.S. and Canada since 1945.

Our road test of the GT6 (April 1967) gave the car high marks: "We approach any car with conventional swing axles with a little apprehension but we found that the GT6 could not be faulted on its handling." Also, ". . . one gets the feeling that the car has a degree of oversteer that can be enjoyed and utilized by a moderately skilled driver while never crossing up an unskilled one. The GT6 corners flat, too, and doesn't seem to want to lift its inside rear wheel in violent low-speed maneuvers."

The GT6 also impressed us with many of its other features and our conclusions about the car were quite enthusiastic:

"In summary, the GT6 is a smaller package that incorporates many of the same qualities that make the Jaguar E-Type such an exhilarating car. It is smooth; it has good torque, low noise level and agility as well as stability in its handling. It's a great improvement over the Spitfire 4 from which it descended. Not that the Spitfire 4 was bad, it's just that the GT6 is so much better. It has no parallel and it's worth the money."

So, the GT6 was quite a pleasant car, in our opinion, and certainly worth the list price of $3039. There was an optional Laycock de Normanville electric overdrive available from the car's inception, and U.S. models of the GT6 came with wire wheels and Dunlop SP-41 radial-ply tires as standard equipment. The GT6 Mark 1 continued in production until 1968, with nearly 16,000 examples being built during its 2-year life span.

The GT6+ (that was the U.S. moniker; in England it was known as the Mark 2) was built from 1968 until 1970 and the total production was just over 12,000 units. The new model featured increased performance and a different rear suspension system. The 2-liter engine was bumped up to 104 gross bhp at 5300 rpm (instead of the previous 95 at 5000) through installation of a new camshaft with more lift and greater overlap, plus slightly larger intake and exhaust valves. Another change, not for performance so much as efficiency, was replacement of the generator by an alternator.

The new rear suspension retained the transverse leaf spring but replaced the swing axle with double-jointed shafts that were located by the spring at the top, a wide-based lower arm and a radius arm. The new arrangement greatly diminished the camber changes and unsettling jacking and tuck-under characteristics of the earlier model. Another alteration worthy of note was a reduction in the seat height to provide more head room for taller drivers who had found the original GT6 uncomfortable at best. The GT6+ models also featured a full range of U.S. safety equipment of the period and raised bumpers both front and rear that did little for the styling but were considerably more practical. And, amazingly, the list price had risen only $6 (to $3045) over the original GT6 two years earlier. The trick to this was the replacement of the previously standard wire wheels with less expensive steel discs.

Our summary of the GT6+ (February 1969) was a query:

GT6+ had revised rear suspension and increased performance.

BRIEF SPECIFICATIONS

	Spitfire 4 & Mark 2	Mark 3	GT6	GT6+
Curb weight, lb	1555*	1680	1970	1975
Wheelbase, in.	83.0	83.0	83.0	83.0
Track, f/r	49.0/48.0	49.0/48.0	49.0/48.0	49.0/49.0
Length	145.0	147.0	145.0	147.0
Width	57.0	57.0	57.0	57.0
Height	47.5	47.5	47.0	47.0
Fuel capacity, U.S. gal.	10.2	9.9	11.7	11.7
Engine type	ohv 4-cyl	ohv 4-cyl	ohv 6-cyl	ohv 6-cyl
Bore x stroke, in.	2.73 x 2.99	2.90 x 2.99	2.94 x 2.99	2.94 x 2.99
Displacement, cc	1147	1296	1998	1998
Horsepower (gross)	63*	75	95	95

*curb weight for Mark 2 is 1630 lb; horsepower is 67.

PERFORMANCE DATA
From Contemporary Tests

	Spitfire 4	Mark 2	Mark 3	GT6	GT6+
0-60 mph, sec	15.5	15.0	13.6	12.3	11.0
Standing ¼ mi, sec	20.8	20.4	19.3	18.8	18.0
Avg fuel consumption, mpg	30.0	27.0	23.0	24.0	24.0
Road test date	4-63	6-65	9-67	4-67	2-69

Engine accessibility was a highlight of Spitfire and GT6 models. The entire front-end sheet metal lifts up to expose engine and front suspension.

"Where else can you get a 6-cyl, 100+-mph coupe with a proper chassis, good finish and jazzy looks for $3000? Nowhere *we* know of."

Buying a Used Spitfire—What to Look for

In conversations with Spitfire owners, including our Engineering Editor who bought a 1964 model in partnership with his brother and which he still owns, it appears there are very few individual quirks to keep in mind when searching for one to buy. The rugged 4-cyl engine (and the 6-cyl from the GT6) is dependable and durable in the manner of British sports car engines of the era. One expert we talked to said he couldn't think of any special engine weak points to check, and after further thought added, "You should look for oil leaks around the timing cover seal and things like that, but no more so than with any used car you'd think about buying."

One of the points that did come up is the transmission tunnel cover which is fiber rather than metal and which can wear out and crack. This will allow noise and fumes into the cockpit and may make you think the gearbox is bad because of the noise level. Because the Spitfires covered here did not have synchro on 1st gear, the prospective buyer should be aware that an unskilled driver might have damaged 1st gear, so listen carefully for undue noise while driving the car in that gear. Also, a Spitfire that has been raced may leak oil from the gearbox tailshaft housing, but that condition rarely occurs with a car that has not been campaigned.

Another item that deserves careful inspection is the front ball joints. The grease boots there can wear out and leave no reservoir of lubricant for the ball joints. While we're at the front of the car, we should mention that the hinges for the large hood/fender section can come loose but don't normally wear out or break. There are rubber bumpers at the junction of the hood and cowl that serve to keep the hood assembly from rattling and these can come off, giving you the impression of an unsound, vibrating front end.

At the rear end, there can sometimes be a clunking noise that would indicate the U-joints between the swing axle and the differential are worn and should be replaced. Looking underneath the car to check for rust coming out of the U-joints is an indicator that this problem will crop up soon if it hasn't already. Otherwise, if the engine, gearbox and differential are quiet, that's a good indication that the car is in good shape as the Spitfire drivetrain components are rated quite durable.

We don't care to go into a long dissertation on the shortcomings of Lucas electrical parts, but the buyer should be aware that these can cause some problems in older cars. Also, one man who is familiar with Spitfires said that generator brushes seem to wear out every 25,000 miles or so with regularity.

Because of the design, the Spitfire and GT6 don't seem to be as prone to rust damage as other British sports cars, especially along the rocker panels. Apparently the body and the double Y-type frame result in fewer nooks and crannies for trapping water, although floor rust in front of the seats can be a problem with the convertible if the carpeting gets wet frequently and is neglected.

Though not a problem area to look for, the prospective purchaser of a Spitfire should give a lot of thought to suspension modifications if he or she is planning to engage in spirited driving, racing or slalom activities. For normal touring, as we said, the Spitfire handles quite well and the independent suspension soaks up bumps and uneven surfaces with efficiency. But, the swing-axle rear suspension does have its limitations. Replacing the shock absorbers with stiffer units (such as Konis), adding a camber compensator (British Leyland used to stock this as a competition part but they don't any longer), re-arcing the rear transverse leaf spring, or putting on the transverse spring (known as a swing spring) from the Mark 4 Spitfire models (1973-on) will make a big difference. Although the Mark 4s have a wider track, the swing spring will fit earlier models with the narrower track.

Conclusion

When it was introduced in late 1962, the Triumph Spitfire impressed enthusiasts with its performance and ride, as both qualities were superior to those of the competitively priced cars. Many sports car aficionados were exposed to the delights of this type of motoring for the first time and then moved on to more sophisticated, and more expensive, cars. In today's world where many of us are going back to basics in our life style, the Spitfire serves as a reminder of why we became enthusiasts in the first place and driving one again after a several-year interval is a refreshing experience. With present-day sports cars becoming increasingly more expensive and the number of true convertible sports cars dwindling, the Spitfire offers an inexpensive alternative capable of delivering ample driving fun. And the GT6, in a different vein, can provide an equal amount of driving pleasure for the sports/GT car buyer.

TYPICAL ASKING PRICES

Year & Type	Price Range
1962-1965 Spitfire 4	$450-800
1965-1967 Mark 2	$600-1100
1967-1970 Mark 3	$800-1500
1966-1968 GT6	$900-1500
1968-1970 GT6+	$1100-1900

Triumph GT6

Above: The 1966 Mk 1 version looking very like the Spitfire on which it was based. Right: Also a Mk 1 but with optional wire wheels. Below: Mark 2 version can be identified by a higher bumper line, louvred bonnet and different wheel trims. Bottom: Mk 3 version had a number of styling and detail changes.

WE NEED the expression Sports Hatch to describe a growing number of cars these days, but 10 years ago the class was not as common. Nowadays we usually quote Ford's Capri II as the obvious example, and there is a lot of choice. In the 1960s the advantages of a "Hatch" had yet to be understood.

Ten years ago if you were fairly rich you bought a Jaguar E-type, but most people happily settled for an MGB GT. The alternative to that — £65 cheaper in basic price but rather smaller — was the sleek little fastback Triumph GT6.

The GT6 was never one of Triumph's front-line stars. In concept, it evolved almost by accident, and it was usually overshadowed by a more hairy-chested Triumph TR4A, TR5 and latterly TR6. Introduced late in 1966, and in production until the end of 1973, more than 40,000 GT6s were sold (of which only about 8,000 stayed in Britain). To some motorists the GT6 was the best thing Triumph had ever done — they were the sort who wanted performance and looks, but not space — to others it was outstandingly impractical.

What was a GT6?

A GT6 was *small*. We must make that quite clear from the outset. It was precisely the same length and width as a Spitfire — much of its body and all of its chassis was shared with the Spitfire — which means that, by definition, it was a strict two-seater. It bore the same basic relationship to a Spitfire as a Vitesse did to a Herald, and the comparison is apt for the mechanical transplant was almost identical.

All GT6s had fastback coupé shells with a big upward-opening loading door/hatch. All had up-rated 1,998 c.c. six-cylinder engines backed by high geared (and close ratio'd) transmissions. All put an emphasis on silky smoothness at the expense of outright sporting handling.

All GT6s had much more storage space than the average for sports cars — but significantly less than that of the MGB GT for instance. Not even athletic children could be carried behind the front seats, as there was absolutely no provision for them.

All GT6s were smart, smooth, somehow sexy, but they were definitely an indulgence. No one who needed to carry more than one passenger — ever — bought a GT6. A GT6 cutomer wanted a sports car without the cold winds, but he also wanted a Jaguar E Type without the expense. For at first, in looks and even — somehow — even in its engine noise, a GT6 was like a mini-E Type. But it could not compare, especially in value for money.

How the GT6 evolved

In the beginning, there should, simply, have been a Spitfire GT; a four-cylinder chassis might have been mated with this fastback body. But in prototype form it was too heavy, too slow, and looked like being too expensive to sell. Very quickly, therefore (Triumph were most accomplished at mechanical conjuring tricks in the 1960s), the chassis was modified, Vitesse-style, to take a six-cylinder engine.

The original GT6 had a tuned-up 1,998 c.c. Triumph 2000 engine with 95 bhp at 5,000 rpm. Behind it was a close-ratio all-synchromesh gearbox (already proved by Spitfire rally cars and at Le Mans), an optional overdrive, and a new chassis-mounted differential. Without overdrive the gearing was very high — top gear ratio was 3.27 to 1 — but a more normal 3.89 final drive was fitted with the optional overdrive. There was never an automatic transmission option.

Front disc brakes were standard, as were radial ply tyres (these were by no means common, even in 1966), and wire spoke wheels were optional. Both wire wheels and overdrive were popular fitments, even though they added nine per cent to the car's price when new. The suspension, unfortunately, was like the Spitfire, with swing axle independent rear layout, but with much softer damping to suit the ride-conscious American market.

The original car, therefore, suffered in our hands because we thoroughly disliked the handling (the swing axles could, and often did, jack up under hard cornering, which produced strong oversteer), we thought the steering too vague, and the ventilation poor. At the conclusion of our test we said: "Potentially the GT6 is a fine formula; with further development (and if necessary a price increase) it could become outstanding."

There were considerable differences in appearance between the Mk 1 (left) and Mk 3 (below). The lift-up tailgate gives the impression of more luggage space than there really was. Spare wheel and tools plus fuel tank lived under the "boot" floor. There were interior changes to seats and facia between the marks too

From the autumn of 1968 Triumph rectified these shortcomings, and the GT6 Mk II was a vastly improved car. It benefited from the new full-width cylinder head developed for the TR5 and 2.5PI (104 bhp at 5,300 rpm), got face-level ventilation and one-way outlets near the tail, and other details. Of paramount importance, though, was its new suspension. This was almost GP-style, with reversed bottom wishbones helping the transverse leaf spring to locate the back wheels properly, and with a flexible rubber "doughnut" in the drive shafts to make all geometry changes possible. Like the earlier cars, too, the ultra-high gearing made good fuel economy possible; 28 mpg was usual, and 33 mpg at a steady 70 mph was also measured.

Two years later (strangely enough the new car was introduced immediately *after* the Earls Court Show of 1970) the Mk II gave way to

Performance Data

	GT6 Mk I o/d	GT6 Mk II o/d	GT6 Mk III
Road tested in *Autocar* of:	7 Sept 1967	3April 1969	23 Sept 1971
Mean maximum speed (mph)	106*	107*	112
Acceleration (sec)			
0-30mph	3.6	3.5	3.9
0-40	5.9	5.2	5.5
0-50	8.5	7.2	7.7
0-60	12.0	10.0	10.1
0-70	15.6	13.7	14.0
0-80	21.2	18.2	18.4
0-90	31.3	26.2	24.5
0-100	—	39.3	35.8
Standing ¼-mile (sec)	18.5	17.3	17.4
Direct top gear (sec)			
10-30mph	—	7.8	—
20-40	6.7	6.5	8.6
30-50	6.9	6.7	7.8
40-60	7.9	6.9	8.3
50-70	8.5	7.9	9.2
60-80	10.0	9.5	10.1
70-90	16.1	13.2	11.7
80-100	—	—	16.9
Overall mpg	20.2	25.2	27.6
Typical mpg	24.0	28.0	28.0
Dimensions			
Length	12ft 1in.	12ft 1in.	12ft 5in.
Width	4ft 9in.	4ft 9in.	4ft 10.5in.
Height	3ft 11in.	3ft 11in.	3ft 11in.
Unladen weight (cwt)	17.5	17.8	18.1

*In overdrive top gear

Milestones

	Series	Chassis No.
October 1966: GT6 Coupé introduced, using basic Spitfire chassis and body, with six-cylinder, 2-litre engine, and fastback coupé hardtop. Two seats, all-independent suspension, front disc brakes, radial ply tyres. Opening chassis number:	KC	1
September 1968: GT6 Mk II replaced original Mk I car. Basically similar, but with 104bhp instead of 95bhp. 'Lower wishbone' type of rear suspension replaced swing axles. Style changes including raised bumper, and revised facia. Final chassis number (Mk I):	KC	13752
Opening chassis number (Mk II):	KC	50001
October 1969: Uprating, including reclining seats and minor style changes, from:	KC	75031
October 1970: GT6 Mk III replaced Mk II car. Similar shape, but mostly new skin panels, including cut off tail, new lights, new bumpers. Final chassis number (Mk II):	KC	832908
Opening chassis number (Mk III):	KE	1
February 1973: Revised Mk III including simpler rear suspension, by swing axles but with "swing spring" feature like Spitfire IV. Nylon covered seats, re-styled instruments. From:	KE	20001
December 1973: GT6 Mk III discontinued at:	KE	24218

Note: *Chassis number sequence misses out block of numbers, but start and finish points were as stated.*

Spares prices

	Mk I	Mk II	Mk III
Engine assembly — bare	£554.04	£319.68	£372.06
Gearbox assembly (exchange)	n.a.	£92.34	£102.06
Clutch cover assembly	£15.11	£15.11	£14.26
Clutch driven plate	£13.44	£13.44	£13.44
Prop shaft U/J repair kit (each)	£4.10	£4.10	£4.10
Final-drive assembly — new	£286.20	£286.20	£286.20
— exchange	£97.52	£97.52	£115.56
Brake pads — front (set)	£11.65	£11.65	£11.65
Brake shoes — rear (set)	£3.51	£3.51	£3.51
Suspension dampers — front (pair)	£23.12	£27.22	£19.22
Suspension dampers — rear (pair)	£24.40	£24.82	£24.84
Radiator assembly	£89.10	£89.10	£54.09
Dynamo (Mk I), alternator (Mks II and III)	£20.77	£26.78	£29.16
Starter motor	£14.22	£14.22	£14.22
Front "wing" panel	£10.58	£10.91	£14.63
Front bumper	£23.49	£27.27	£40.50
Rear bumper (both sides)	£16.85	£13.72	£42.39
Windscreen, toughened	£15.88	£15.88	£15.88
Windscreen, laminated	£23.49	£23.49	£23.49
Exhaust system, complete	£25.92	£32.89	£47.74

(All the above prices include VAT at 8 per cent)

the Mk III. This was almost entirely a re-styling job, with a more smoothly-shaped bonnet, and a cut-back tail having much more integrated panel/lamps/bumper arrangements. It was done to commonise it with the similarly re-shaped Spitfire, announced at the same time. All the louvres and grilles introduced earlier were eliminated, and the Mk III looked the smoothest of all GT6s. There were many other practical improvements, including a revised facia. By then, too, a GT6 was really quick, with 112 mph maximum speed capability, but it still retained the remarkable fuel consumption possibilities.

The Mk III was built until the end of 1973, when burgeoning North American safety legislation finally killed it off. (The Spitfire, of course, is still with us.) There was one more important change; in February 1973 the "lower wishbone" rear suspension was dropped in favour of the simpler, if not quite as effective, "pivoting spring" system adopted in 1970 by the Spitfire IV. Only about 4,000 cars (most of which were exported) were built with this final chassis layout.

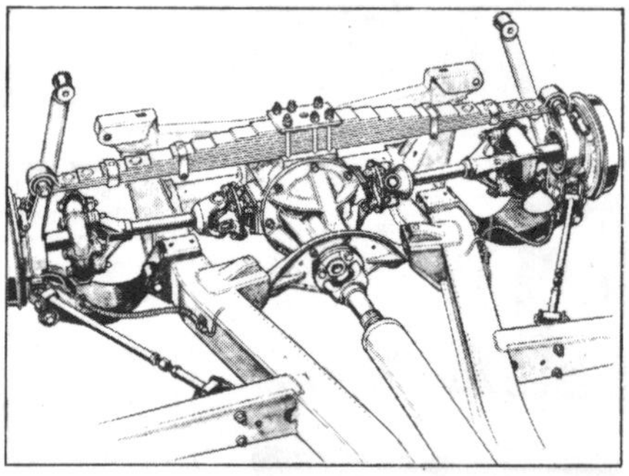

Above: Like the Spitfire and Herald, engine accessibility was good with the whole bonnet swinging forward out of the way. This is a Mk 2 but looks very similar to Mk 1 or Mk 3 versions. The 1,998 cc engine is a six-cylinder. Left: The interior of the original Mk 3. Below: This revised rear suspension was introduced for the Mk 2 and gave much better rear wheel location

What to look for

As all GT6s used the same basic chassis, engine and transmission, the choice has to be between body variations and suspensions. Let us deal, first, with suspensions.

We always treated a Mk I GT6's handling with the gravest suspicion, and not even Triumph would now argue with us on that. Nearly 2,700 GT6s of this type were sold in Britain, the youngest will now be more than nine years old, and we would not frankly recommend it to you, unless you wanted to spend only a little money.

Any other GT6, built from September 1968, is a thoroughly satisfying little car. The "lower wishbone" rear suspension is slightly more effective at the limit than a "pivoting spring" suspension, but we think this is not significant.

Next there is the question of gearing and transmissions. Many GT6s were built with the optional overdrive (and remember that an update is not too difficult if you know your Triumphs and you find a GT6 without overdrive), and we think it did much for the car. Normally overdrive came with a 3.89 differential, but certainly in Mk II days quite a few cars were delivered with overdrives *and* 3.27 axles; that gave astonishingly high gearing (about 25 mph/1,000 rpm in overdrive top) which suited some people. [illegible] is a desirable fitting, if you are shopping around in the secondhand market.

Some GT6s had centre lock wire wheels. These are very pretty, and quickly changed in case of puncture (but how often do you get punctures nowadays?), but the very devil to keep clean, true, and in good condition. We prefer the disc wheels, particularly those fitted to Mk IIIs, which were both practical and attractively styled.

At one time, Triumph listed an "occasional rear seat" for the GT6. Don't grieve if you can't find a car with one fitted, as it was entirely impractical. We stick to our guns, and call *all* GT6s as strict two-seaters.

The later cars — Mk IIIs in particular — had nicer trim and more complete equipment. We prefer the Mk III's uncluttered exterior detailing, and its lack of lumps, bumps, louvres, and grilles. All GT6s built after the autumn of 1969 had partly reclinable seats, which is nice if you are long in the arm; as sleeping-type recliners they were not very useful, but the ability to alter your driving position is always welcome. Some GT6s were built with laminated screens which is a nice extra if you are safety conscious.

What about rust and general deterioration? The GT6 is not at all out of the ordinary, with its pressed-steel body, though it is always nice to know that there is a separate box section chassis frame (laid out in the shape of a double-backbone, and very robust). It means that a GT6, even in poor visual condition, should be structurally sound, or reclaimable.

A badly rusted body will betray itself by its sills and by the boot floor (under the spare wheel, and out of sight on this model). The first important structural problem will be when the radius arm pick-up points in the body shell (behind the seats in the floor pan) begin to break up. These *can* be repaired, but need careful attention to preserve the suspension geometry.

If parts of the bonnet are in a poor condition, remember that the entire swivelling assembly has to be replaced, and that this could be expensive; getting the settings right, too, may be finicky.

Look carefully for crash damage on the chassis. Though it is strong for its "as designed" job, it may have suffered suspension pick-up derangements at one time, and the members might have been distorted. Bushes and trunnions, in general, may be worn (anything over 25,000 miles is a good life for suspension pivots) — most are very simple to replace, and spares are readily available. The steering should be slightly flexible on its mountings (that is what the rubber pads are for), but these may get oil-soaked and may lead to rack float, which plays havoc with the steering's precision.

In this car, the big 2-litre engine's job is simple. Although it is relatively powerful it looks after a light car, and seems to last very well. Engine compartment accessibility is superb — probably the best of any relatively modern car. Remember that if necessary you can take the entire bonnet off, and find even more space for work on the front end of the car.

The transmission seems to stand up well, though the earlier models seem to wear out their synchromesh rather early. Later boxes are better in this respect. Look at the rear suspension "doughnuts" from time to time for tearing and general old age. They are cheap to buy, and simple to replace; as a "safety valve" to the rest of the drive line they deserve to be kept in the best condition.

Lastly, the spares situation on GT6s *at the moment* is quite satisfactory, particularly as so many parts are shared with the Spitfire and to a lesser degree with the Vitesse (which is also obsolete). However, Leyland seem to lose interest rapidly in cars like the GT6 which do not fit into their mass-production plans, and it could be that within a couple of years unique parts might become a rarity. Trim and decoration details will be the first to suffer. You have been warned.

Approximate selling prices

Price Range	GT6 Mk I	GT6 Mk II	GT6 Mk III
£400-£450	1967		
£500-£550	1968		
£650-£700		1969	
£750-£800		1970	
£950-£1,000			1971
£1,100-£1,200			1972
£1,300-£1,400			1973
£1,600-£1,700			1974

Note: *Overdrive was always optional on the GT6, many cars were so fitted, and it may increase the value of a secondhand car by up to £50*

SPARES FOR ALL GT6 MODELS!
John Hill's Dept GT6
British Leyland Heritage Approved Suppliers
ARTHUR STREET, REDDITCH, WORCS.
B98 8JY
(STATE MODEL) FOR
CATALOGUE AVAILABLE
FOR
COMPREHENSIVE DETAILS
NEW RECONDITIONED – SECONDHAND
Tel: Redditch (0527) 20880
Seat recovering, carpets, headrests, door & other trim casings, bumpers, lights, grilles, valances & badges.
Spoilers, sills, door skins, wings, full selection of repair palens.
Suspension overhaul kits, springs & dampers, brake discs & overhaul kits.
Exhausts & clamps.